THE LITTLE BOOK OF
TREES

THE LITTLE BOOK OF
TREES

With color illustrations by Tugce Okay

HERMAN SHUGART AND
PETER WHITE

PRINCETON UNIVERSITY PRESS
PRINCETON AND OXFORD

Published in 2024 by Princeton University Press
41 William Street, Princeton, New Jersey 08540
99 Banbury Road, Oxford OX2 6JX
press.princeton.edu

Copyright © 2024 by UniPress Books Limited
www.unipressbooks.com

All rights reserved. No part of this book may be reproduced or transmitted
in any form or by any means, electronic or mechanical, including photocopying,
recording, or by any information storage-and-retrieval system, without written
permission from the copyright holder. Requests for permission to reproduce
material from this work should be sent to permissions@press.princeton.edu

Library of Congress Control Number 2023943763
ISBN 978-0-691-25179-0
Ebook ISBN 978-0-691-25180-6

Typeset in Calluna and Futura PT

Printed and bound in China
1 3 5 7 9 10 8 6 4 2

British Library Cataloging-in-Publication Data is available

This book was conceived, designed, and produced by UniPress Books Limited

Publisher: Nigel Browning
Managing editor: Slav Todorov
Project development and management: Ruth Patrick
Design and art direction: Lindsey Johns
Copy editor: Caroline West
Proofreader: Robin Pridy
Color illustrations: Tugce Okay
Line illustrations: Ian Durneen

IMAGE CREDITS
Alamy Stock Photo: 17 Kevin Schafer; 48 Trent Dietsche;
109 Michael Gadomski; 135 Douglas Peebles Photography;
149 Panoramic Images. **Dreamstime.com**: 10 Serghei Ciornii;
80 © Sylvie Lebchek; 98 © Laurentius Pattyn; 142 © Photoholidaysjk.
iStock: 33 mantaphoto; 57 Iamraffnovais; 118 MendelPerkinsPhotography.
Nature Picture Library: 125 SCOTLAND: The Big Picture. **Shutterstock**:
23 sittitap; 38 Jadetanadee; 44 Tomas Vynikal; 61 Birds Fish Trees;
71 Rejdan; 87 Malene Thyssen; 90 Emilio100; 101 Zyankarlo; 141 sduraku.
Other: 28 Graph data: Jwratner1 Wikipedia; 67 Robert Kerton CSIRO;
69 NC State Extension Publications; 114 Nartreb www.davidalbeck.com.
Additional illustration references: 107 Brocken Inaglory.

Also available in this series:

THE LITTLE BOOK OF
BEETLES

THE LITTLE BOOK OF
SPIDERS

THE LITTLE BOOK OF
BUTTERFLIES

Coming soon:

THE LITTLE BOOK OF
FUNGI

THE LITTLE BOOK OF
WEATHER

THE LITTLE BOOK OF
DINOSAURS

THE LITTLE BOOK OF
WHALES

CONTENTS

INTRODUCTION

Independently and early in our careers, both of your authors enjoyed the remarkably diverse forests of the Great Smoky Mountains National Park, in the United States—an ecological celebration of trees growing, trees forming forests, forests sheltering self-generated gardens of wildflowers, forests replicating continental patterns up and down the mountain slopes. With trees, forests, and forested landscapes, the more one knows about them, the more remarkable they become. This little book relates the wonder and beauty of trees and the forests that they self-generate.

THE POWER OF TREES

That the geographical patterns of forests and the forms of trees can be related to patterns in the environment was documented by the 4th century BCE Greek philosopher, Theophrastus, in his *Enquiry into Plants*. Nowadays, these regularities in pattern look for explanations in the evolutionary adaptations of trees, or in the economics of trees' patterns in allocating their resources in optimal ways. Any compendium on trees presents difficult riddles on the "Why?" of the interrelatedness of forms, and patterns. A species of tree can be deciduous, losing all its leaves annually. A close relative may be an "evergreen" tree, which gains new leaves before losing old ones, thus remaining green all year round. Both species may share an evolutionary relative that is not a tree at all—rather a vine, or a shrub, or an herbaceous plant. Conversely, a collection of deciduous trees, even when their leaves are of similar shapes and sizes, may hardly be evolutionarily related at all.

Humankind has its origins in the savannahs of Africa and understanding the flowering and fruiting of trees formed part of our survival knowledge. Indeed, a knowledge of trees and their attributes is an essential survival skill in the development of human culture. Even as we finished writing this book, in May 2023, an amazing story of four children, aged 13, 9, 4, and 11 months, living for 40 days in Colombian rainforest as the sole survivors of a light aircraft crash has been documented. The children were members of the indigenous Huitoto people, and the older ones had survival knowledge that saved them all. Clearly, the potential power of an indigenous knowledge of forests and trees cannot be overestimated—it certainly wasn't lost on our ancestors.

WHAT THIS BOOK COVERS

There is much to be discussed in this little book. We start by defining what a tree is: generally a tall, single-stemmed plant that grows from its branch tips. But there are striking variations in "How tall is tall?" or "Just how tall can trees be?" The tree life-form is old. It develops in club mosses, redevelops in tree ferns, gymnosperm trees, and angiosperm trees . . . One of the odd aspects of the invention and reinvention of the tree life-form is that tree roots are more recent, if one considers almost 400 million years ago recent. The development of roots allowed trees to spread over the terrestrial surface of the Earth and may have changed the planet's chemistry. The globalization of the tree life-form has since given rise to current patterns in tree diversity, with the tropical rainforests as the apogee of this diversity.

The range of patterns found in leaves, tree trunks, bark, and roots indicates the variations in the tissues of trees and tree function. They also come together to provide a basis for the overall geometry of trees. In a similar manner, population change and reproduction in trees produce patterns in space and time in forests that are assembled from these trees. It is our hope that the *Little Book of Trees* will ignite your interest in these impressive plants and inspire you to help conserve them against the damaging effects of climate change, pollution, overharvesting, and habitat loss.

Herman Shugart and Peter White

WHAT IS A TREE?

A tree is a tall, woody, perennial plant with an elongated stem or trunk. The trunk typically supports woody branches, twigs, and leaves. "Tree" is not a taxonomic grouping; it is a life-form of plants with tree attributes. Some trees are evolutionarily related with shared ancestors. However, other trees have independent origins, even though they seem similar. One might expect the nearest relatives of a tree to be other trees, but this is not always so. There are many plant genera with tree, shrub, and/or vine life-forms that are strongly related from an evolutionary point of view. DNA analyses provide multiple examples of trees with kindred species of alternative plant life-forms. An extreme case is the large plane trees of the genus *Platanus*, which are most related to the aquatic sacred lotus (*Nelumbo nucifera*) and vice versa. Plane trees are also the lotus's nearest kin. Their last common ancestor dates from 60 Mya (Mya = million years ago).

COMPETITION BETWEEN PLANTS

A tree's height gives it preemptive access to incoming light, a winning strategy when other resources (water, soil nutrients) are in ample supply. Light competition is one-way: the taller plants get sunlight first and the smaller plants can only access the overhead light that has streamed past the taller plants above. In contrast, competition between individual plants for nutrients and/or water is two-way—when a tree's roots capture nutrients or water, these are then made unavailable to other plants, regardless of their sizes.

ADVANTAGES AND
DISADVANTAGES OF HEIGHT

There are many advantages to height among plants competing for resources, but there are also challenges in being a tall plant. Transporting water and nutrients to leaves high in the air requires roots to supply water transport systems in the trunks of trees. In general, plants require some sort of woody fiber to attain height. In most trees, this involves two different modes of growth: primary growth (extending roots and branches) and secondary growth (thickening the trunk, branches, and roots). Along with primary and secondary growth, the evolution of tree-like plants has also involved solutions to several structural problems—forming self-supporting structures, monopodial trunks, and fluid transport in tall plants—all of which are discussed in this chapter.

← The Oriental plane tree (*Platanus orientalis*). The nearest living relative to plane trees (*Platanus*) is the aquatic and herbaceous sacred lotus (*Nelumbo*).

THE TREE LIFE-FORM

Evolutionary relatedness of plants traditionally involved determining fossil dates and comparing morphologies, particularly of flowers. Modern DNA analysis is revealing surprising relations among plants. Flowering plants (angiosperms), originally thought to have arisen from a single ancestor, may actually be a collection of species with five different origins. The "clock" provided by the accumulation of DNA mutations has pushed separations between taxa back as much as 60 million years. Microscopic detection of fossil pollen has displaced the origins of different seed plants by a similar degree.

The tree life-form developed in different evolutionary pathways as solutions to the problems of building plants with tall, stable trunks. The trunks of trees can be formed in several different ways, depending on the structural challenges involved in building taller plants (wood macro-geometry problem); growing the wooden parts of trees to adjust for increasing the tree's height and mass (secondary-growth problem); and lifting water to the treetops (hydraulic problem).

↓ Female cones of the monkey puzzle tree (*Araucaria araucana*) are round, 4½–8 in (12–20 cm) across, and hold about 200 seeds.

↓ Male (pollen) cones are initially 1½ in (4 cm) long. They expand to 3–4½ in (8–12 cm) long by 2–2½in (5–6 cm) wide at pollen release.

→ *Araucaria araucana* trees growing in Conguillío National Park, in the Lakes District of Chile. Monkey puzzle trees share a common ancestry with trees from the time when Australia, Antarctica (fossil plants), and South America were linked by land in the supercontinent known as Gondwanaland.

NON-SEED TREES OF THE FIRST FOREST

Trees and forests originated 393–383 Mya in the Middle Devonian Period. In 2020, Professor William Stein and colleagues, of Binghampton University in New York state, excavated the oldest such intact forest in the Catskill region near Cairo, New York. They uncovered and mapped a buried ancient soil and fossilized tree roots across an area measuring ¾ acre (3,000 m²). This ancient forest existed at the transition of the Earth from a planet virtually devoid of forests to one covered by them. Professor Stein and his colleagues' detailed mapping of this area allows us to visualize a forest that is now lost deep in time.

EARLY FOREST PLANTS

The trees of early forests reproduced through spores, not seeds. The roots and the size of the trees at the Catskill site help us imagine what this incredibly ancient forest must have looked like. There were a few *Stigmaria*, the roots of lycopsids—the so-called "giant club mosses" that could reach heights of 98 ft (30 m). The forest was also home to tree-fern-like trees called cladoxylopsids, which have no living relatives but are related to ferns and horsetails (*Equisetum*). At full size, these stood up to 40 ft (12 m) tall and had no leaves, but they did develop a canopy of photosynthesizing twigs and branches. Unusually detailed fossil cladoxylopsids from China indicate that they were hollow, and as the plant grew larger, the resulting strain would tear the trunk. These rips and tears would repair themselves over time to produce diameter growth—a violent, Procrustean solution to the secondary-growth challenge of increasing the diameter of the trunk as a tree grows taller.

THE *ARCHAEOPTERIS* GENUS

At the New York site, protogymnosperms of the genus *Archaeopteris*, distant ancestors of modern gymnosperms, were abundant. These trees could grow up to 100 ft (30 m) tall and had base diameters of almost 5 ft (1.5 m). The name *Archaeopteris* derives from the

Greek *archaîos* (ancient) and *ptéris* (fern), and the plants were initially classified as ferns. They had leafy, fern-like foliage arranged in umbrellas of fronds. The resultant canopy captured light efficiently and the leaves may have been shed seasonally, as in modern deciduous trees. The roots resembled those of today's trees in that they were both spatially extensive and deep. These then-novel roots provided the buried organic carbon compounds that could change the chemistry of soils.

The resounding success of *Archaeopteris* covered the planet with a soil-building, biomass-accumulating forest unlike anything that had come before. Storing carbon in trees and forest soils, it contributed to the chemistry of the planetary atmosphere and the oceans, much like modern forests.

↑ *Archaeopteris* trees were the first to evolve deep and extensive root systems like those of today's trees, leading to the building of soils and generating planetary chemical changes.

ARCHAEOPTERIS DOMINATION

Archaeopteris trees came to dominate the Earth and eventually composed 90 percent of the planetary forests. The genus remained until the Lower Carboniferous, about 50 million years after the time of the early forests mapped near Cairo, New York. *Archaeopteris* fossils have been reported from North and South America, Europe, and Asia. One species, *Archaeopteris notosaria*, grew within the then Antarctic Circle and fossilized *Archaeopteris* fronds have also been found on Bear Island above the Arctic Circle in the Norwegian Svalbard archipelago.

DEVONIAN TREES

About 372 Mya, the Devonian Epoch ended with a massive and complex extinction, particularly of the remarkably diverse coastal reefs, primarily in tropical, shallow water, which had been evolving and diversifying until that time. This extinction was one of the "Big Five" such occurrences in the Earth's history.

The Late Devonian extinction appears to be the result of several different events. The first of these, 372 Mya, was the Kellwasser Event, a massive extinction pulse of marine invertebrates. Evidence of this event can be seen in the rock strata in the Kellwasser Valley in Lower Saxony, Germany. In these strata, the rocks contain two distinct shale layers, which implies that there were two occurrences of ocean anoxia (lack of oxygen). The Hangenberg Event, which occurred 13 million years later and was named for strata from the German Hangenberg Black Shale, documents an extinction event associated with a disappearance of reefs and many fish genera.

There is little debate over whether the collective events of the Late Devonian were anything but horrific. But there are several explanations and considerable debate as to the complex cause(s) of the events of these Late Devonian extinctions. Collectively, they eliminated about 80 percent of the planet's animal species, mostly from the diverse array of marine species of the shallow tropical seas.

PLANETARY EFFECTS
OF FORESTS

Although the causes of the Late Devonian extinctions are complex, with competing potential reasons, initial forestation across Earth might have played a role in massive planetary change. Since we are now in a time of planetary deforestation, appreciating the global effects of forests is central to understanding how Earth functions.

WHAT CAUSED THE
LATE DEVONIAN EXTINCTIONS?

One theory regarding the possible causes behind these extinctions indicts the success of *Archaeopteris* roots in weathering soils, increasing runoff, and generating global changes in tropical ocean chemistry. Professors Meyer-Berthaud, Scheckler, and Wendt have conjectured that the primeval, early success of *Archaeopteris* contributed to a large decrease in carbon dioxide in the atmosphere (from 10 to 1 percent) and an increase in atmospheric oxygen (from 5 to 20 percent)—conditions more like those of today. These chemical changes could have induced the much cooler temperatures involved in the Late Devonian Mass Extinction.

↑ Fossil trilobites from the Devonian Epoch in Ontario, Canada. Trilobites were a diverse, abundant, and dominant animal species in Devonian coastal environments. They were greatly diminished by the end of the Devonian.

HARDWOODS AND SOFTWOODS

Foresters classify trees according to the features of the wood they produce. For practical foresters, trees are either softwoods, which are more easily sawn and nailed, or hardwoods, which are more difficult to saw and join. This utilitarian classification is clouded by the additional formal classification used in forestry that uses "hardwoods" to describe tree-form angiosperms (flowering plants) and "softwoods" for tree-form conifer trees (in the gymnosperm division, Pinophyta).

In combination, these definitions produce glaring paradoxes when comparing hardwoods and softwoods. For example, balsa (*Ochroma pyramidale*) is a hardwood tree in the sense that it is an angiosperm, but it has the softest commercially available wood. The opposite also occurs and some softwoods (conifers) can have very hard wood.

TYPES OF TREE TISSUE

Macroscopically, angiosperm and gymnosperm trees have similar arrangements of plant tissues, seen as rings in cross section, which allow for secondary growth. Both angiosperm and gymnosperm trees are typically covered by bark, which varies greatly from a fleshy, smooth, almost skin-like covering to a very hard and thickened, rough covering. Bark is produced by a thin layer of bark cambium tissue that grows outward to produce bark and repairs the outer protective bark layer.

Inside the bark is a layer of phloem tissue, which primarily transports the sugars made by photosynthesis. Phloem tissues may also contain specialized cells that provide mechanical support, flexibility, storage, and protection from herbivores. Phloem is composed of living tissue in contrast to the dead cells in most of the xylem tissue. Inside the ring of phloem tissue is another ring of living tissue called the vascular cambium. This has two kinds of cells. Some are long and oriented along the growth axes of the tree (fusiform initials) and eventually produce the xylem tubes that transport water to the tree's leaves. Other cambial cells are smaller and produce rays that conduct water outward from the tree's center.

XYLEM IN ANGIOSPERMS AND CONIFERS

Angiosperms and conifers converge at the macro level in forming stronger, tougher, more supportive heartwood over time in the water-conducting sapwood. But they diverge microscopically in how the xylem cells solve the strength versus water-conduction trade-off. In conifers, the xylem cells are tracheids, which transport water upward and provide the mechanical strength for building the tall, wooden tree structure. Thick-walled tracheids provide strength, while thinner-walled tracheids, which are larger in cross-sectional diameter, transport more water, faster. In angiosperms, the xylem is more complex, since the tissues have specialized wood fibers to give the xylem wood mechanical strength. Water-conducting pit vessels are collections of tracheids and other structures that form a wider cross section and allow for greater volume in the water transport system and faster velocities of water moving upward.

The innermost part of the tree consists of xylem tissue produced by the vascular cambium. This tissue is involved in the transport of water and nutrients and, unlike the sap-transporting phloem, is composed of dead cells. Typically, in both angiosperms and conifers, the outer xylem wood is lighter in color and weight and is called sapwood. In most trees, the older xylem found in the tree's center is chemically transformed to form heartwood, which is often darker, heavier, more decay-resistant, and stronger than the younger sapwood surrounding it.

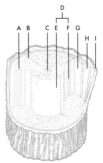

→ Cross section of a tree trunk showing the key structures, as follows: (A) annual ring, (B) rays, (C) pith, (D) xylem, (E) heartwood, (F) sapwood, (G) vascular cambium, (H) phloem, (I) bark.

DEVELOPING WOODINESS ON ISLANDS

Tall plants reaching heights of tens of meters require wood fibers for strength. A conspicuous feature of the plants growing on remote islands is a high proportion of woody plants that evolved from herbaceous continental ancestors. Charles Darwin (1809–1882) noted this tendency during his five-year journey on HMS *Beagle*, and later, his 1859 book *On the Origin of Species* attributed this to natural selection. The phenomenon of insular woodiness can also occur on continents and is referred to as secondary woodiness.

Professor Frederic Lens and colleagues, of Leiden University, tabulated the occurrences of woodiness in plants of herbaceous origins on the Canary Islands. There, they found that 220 native flowering plants represent 34 genera from 15 plant families—this diversity originated from 38 events in the development of woodiness in herbaceous lineages. While there are clearly advantages in the height granted by woodiness over herbaceous competitors, insular woodiness is also associated with drier conditions and drought resistance.

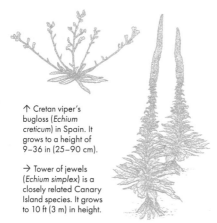

↑ Cretan viper's bugloss (*Echium creticum*) in Spain. It grows to a height of 9–36 in (25–90 cm).

→ Tower of jewels (*Echium simplex*) is a closely related Canary Island species. It grows to 10 ft (3 m) in height.

→ The dragon tree (*Dracaena draco*) is native to the Canary Islands. A member of the Asparagus family, it has secondary woody tissue and can attain heights (and widths) of 50 ft (15 m) in nearly frost-free climates.

GLOBAL DIVERSITY

Forests cover about one-third of the land surface of our planet. In the far north and south and at high elevations, the forest limit (the "tree line") occurs when the length of the growing season decreases to 95 days and the average air temperature in the growing season falls to 43.5°F (6.4°C). Rainfall is also critical, with forests giving way to savannas, grasslands, and deserts in areas with precipitation less than about 30 in (75 cm) in cooler areas to less than about 39 in (100 cm) in tropical areas.

COUNTING THE WORLD'S TREES

In Chapter 1, we learned that a "tree" is a growth form rather than a taxonomic group and that tree characteristics vary continuously. This means that many definitions of "tree" are possible, employing different values of height, diameter, and branching pattern. To estimate the global diversity of trees, a team led by Italian ecologist Professor Roberto Gatti adopted the tree definition published by the International Union for Conservation of Nature (IUCN): "a woody plant with usually a single stem growing to a height of at least two meters, or if multi-stemmed, then at least one vertical stem five centimeters in diameter at breast height." Using this definition, Gatti and his team estimated in 2022 that the global number of tree species was an astounding 73,000, of which fewer than 65,000 have been given scientific names. Most of the unnamed species occur in tropical forests. Given the high diversity of these forests and the incomplete nature of our tree inventory, some scientists have put the number of tree species at more than 100,000.

→ Tropical rainforests contain 75 percent of the world's tree species, including many that have not yet been scientifically named or documented.

ELEVATION AND TREE DIVERSITY

Ascending mountain slopes is like traveling toward higher latitudes. Temperatures generally get colder and tree diversity declines in both cases. In some instances, however, the lowest slopes are so dry that closed forests yield to open woodlands and tree diversity is limited. Also, moisture availability is affected by topographic contours at all elevations. Water drains from convex shapes (ridgelines) and accumulates in concave shapes (valleys); at any single elevation, diversity tends to decrease from more moist to drier sites. Mountains exhibit complex environmental gradients over relatively small areas along which tree composition shifts as environmental conditions change.

TROPICAL RAINFORESTS

The diversity of most groups of organisms increases dramatically from boreal and temperate latitudes to tropical ones. For trees, the latitudinal gradient is striking: the Earth's coldest forests have, collectively, several hundred tree species, the temperate forests have several thousand, and the tropics have over 60,000 species. Thus, some 75 percent of all tree species are tropical, although tropical forest comprises only about 6 percent of the Earth's land surface. Further, many tropical tree species are rare, underscoring both the difficulty of the taxonomic task ahead and the vulnerability of tropical trees to extinction.

Tropical tree diversity is also impressive at smaller scales. One ecologist calculated that there were more trees species in one square kilometer in Borneo than in all the temperate zone. Sample plots of one hectare (100 × 100 m) have been found to contain 200–650 tree species in tropical rainforests, while the richest temperate zone hectare plots have just 25–50 tree species.

↓ Ebony trees in the genus *Diospyros* have dark and very dense wood, which reflects their slow growth rates and long life spans.

↓ In contrast, balsa (*Ochroma pyramidale*), another rainforest tree, has very low-density wood due to fast growth rates and short life spans.

→ Tropical rainforest on Barro Colorado Island in Lake Gatún, Panama. Rainforests support over 200 tree species in 2½-acre (1-hectare) study plots. The leaves of rainforest trees are generally evergreen and lack the "teeth" (jagged edges) common in the temperate zone.

WHY ARE THERE SO MANY SPECIES IN THE TROPICS?

Why does diversity increase as latitude decreases? Here we group answers to this question under three headings: the energy–diversity theory, the time–stability theory, and theories about species interactions that we classify as "diversity begets diversity." Firstly, a note of caution: these theories are hard to test due to the evolutionary time scales and complex species interactions involved. The mechanisms also cannot be easily isolated from one another in nature. Secondly, the explanations are not mutually exclusive—all could play a role.

ENERGY–DIVERSITY THEORY

The energy–diversity theory states that, as ecosystem energy flow—often represented by net primary production (gain of biomass through photosynthesis) or by environmental predictors of primary production (temperature and water supply)—increases, so diversity increases. The energy–diversity theory argues that higher energy flow translates to more ecological opportunity or the greater persistence of tree populations, thus stimulating greater speciation and reducing extinction. Put another way: areas of low energy flow are more stressful. According to this reasoning, ecosystem diversity is limited by the increasing harshness of environmental conditions from low to high latitudes. In other words, species must adapt to tolerate harsh conditions, which is metabolically costly, limiting or slowing species evolution.

THE DIVERSITY OF LIFE

The increase in diversity from the coldest to the warmest environments occurs not just for trees, but for almost all groups of plants and animals. The fossil record shows that this pattern developed early in the history of life and persisted or redeveloped after periods of changing climates and extinction events. •

TIME–STABILITY THEORY

The time-stability theory proposes that the latitudinal gradient in diversity occurs because the long-term stability of environments increases from the poles to the tropics, leading to higher speciation rates and lower extinction rates in the tropics. Further, because high-latitude environments experience greater seasonality and year-to-year variations in environmental conditions, they select for species with wider environmental tolerances, thus preventing the evolution of narrow specialists. This argument does not depend on speciation rates per se. However, if these are higher in the tropics, diversity would result not just from long-term environmental stability, but also because more species arise over a given interval of time as one moves from cold to tropical environments.

BIODIVERSITY BEGETS BIODIVERSITY

The third set of theories we classify as biodiversity begets biodiversity. This proposes that biological interactions are stronger and more specialized in the tropics, stimulating diversification. One example is the Janzen–Connell Hypothesis, which argues that pathogens and other specialized herbivores accumulate under adult trees of a particular species. Thus, seedlings have lower success near the parent tree than some distance away. This limits the abundance of any one species, which explains the rarity of many tropical tree species, and maintains higher overall diversity because this mechanism potentially limits population increases of individual species.

Biological interactions can include other mechanisms, too: for instance, higher diversity may translate to greater competition, leading to further specialization of trees along environmental gradients. Competition in high-diversity environments could lead to diversification in other ways, such as the evolution of a wide range of tree regeneration strategies. Mutualisms may also support higher diversification, including specialized relationships between trees and pollinators, dispersers, and soil fungi.

↓ A germinating seedling is vulnerable to specialized fungal pests if it is too close to large trees of the same species.

TREE DISTRIBUTIONS IN BIOMES

Ecologist Robert Whittaker (1920–1980) classified the Earth's ecosystems into nine biomes and correlated these with annual temperatures and precipitation levels. Five of the biomes are dominated by forests: tropical rainforest, tropical seasonal forest/savanna, temperate deciduous forest, temperate rainforest, and boreal forest. Whittaker's scheme gives us a starting point for understanding tree distributions in an environmental context.

ENVIRONMENTAL FACTORS

The importance of environmental conditions in explaining tree distributions is based on the idea of evolutionary and ecological trade-offs. In other words, being adapted to one set of conditions generally reduces or precludes success in other sets of conditions. Put another way, trees have different, though sometimes overlapping, environmental distributions. Further, some trees have narrow distributions, while others are more tolerant of environmental variation and have wider distributions. However, no trees are unlimited in environmental tolerance and no species are ubiquitous across the varied climatic conditions of the planet. Ecologists now use statistical models based on environmental distributions to predict how climate change will affect forest cover.

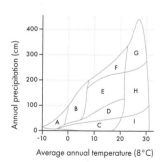

↙ This graph shows the distribution of major biomes as a function of temperature and precipitation.

(A) tundra, (B) boreal forest, (C) temperate grassland/cold desert, (D) woodland/shrubland, (E) temperate seasonal forest, (F) temperate rainforest, (G) tropical rainforest, (H) tropical seasonal forest/savanna, (I) subtropical desert.

28

Whittaker's scheme is generalized at a global scale, but we would find the same concepts at smaller scales. Consider mountain slopes. Generally, temperatures fall and precipitation and cloud cover increase with elevation. Moving on this elevational gradient from warm and dry to cold and wet, species composition changes, as species dependent on longer growing seasons at lower elevations are replaced by cold-adapted species at higher elevations.

OTHER FACTORS TO CONSIDER

We can also learn by considering two factors left out of Whittaker's biome scheme. Firstly, it does not include variation in soil or substrate. Biomes each have a range of soils, from well-drained soils to thin, droughty soils, on the one hand, to inundated and swampy areas on the other. Soils and substrates vary from nutrient rich to nutrient poor, and some forests occur on soils with extreme chemical or physical conditions, like the mangrove forests dominating subtropical and tropical coasts. Temperate conifer forests, except those with high rainfall, are not well represented in the biome classification. Conifer forests are also found in climates with limited summer rainfall, low soil nutrients, and/or frequent fire. Secondly, the scheme focuses on mature forests. Patterns of disturbance, natural or human-caused, and succession are also important to tree adaptations, a subject covered in more detail in Chapter 10.

CONTINENTAL DISTRIBUTION

While tree species are generally confined to particular continents due to evolutionary separation, some genera are characteristic of specific biomes. Thus, in Northern Hemisphere boreal forests, we find spruce, fir, larch, birch, and aspen. Temperate areas include oaks, maples, cherries, ashes, willows, and pines. The tropics and Southern Hemisphere, due to longer evolutionary separation, have less in common, yet some genera are characteristic like figs (*Ficus*) and *Acacia* in tropical forests and southern beech (*Nothofagus*) in temperate forests.

DIVERSITY ANOMALIES

E volution, even in similar environments, has taken independent pathways on different continents, at least before human intervention allowed some species to transcend natural barriers to dispersal such as oceans and mountains ranges. As Charles Darwin wrote in his *On the Origin of Species* (1859):

> . . . *neither the similarity nor the dissimilarity of the inhabitants of various regions can be wholly accounted for by climatal . . . conditions. There is hardly a climate or condition in the Old World which cannot be paralleled in the New . . . [yet] how widely different their organic productions [that is, their species]!*

That similar environments found in different parts of the world have different species profoundly affected Darwin, for it pointed to a cause of species origin that was independent in different places. However, this brings us to another question: while the species lists vary by continent, is the number of species the same for each climate type? Historic biogeography examines the effect of evolutionary history, rather than environments, in shaping distributions.

↓ The bigleaf magnolia (*Magnolia macrophylla*) is native to eastern North America. Its closest relative is found in China's deciduous forests.

↓ Black gum (*Nyssa sylvatica*) trees are found in the wild only in eastern North America and East Asia.

→ Dawn redwoods (*Metasequoia glyptostroboides*) are called "living fossils" because they were known from the fossil record before being discovered in the wild in China in 1941. They were once found across temperate forests but became extinct in all areas apart from East Asia during the repeated glaciations of the Pleistocene Period (5 Mya to 11,000 years ago).

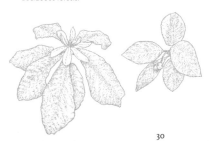

EVOLUTION AND DIVERSITY PATTERNS

In the 1700s, Swedish botanist Carl Linnaeus (1707–1778) sent students all over the world. When specimens were returned, he noticed an odd pattern. For some North American species, the closest relatives were from half a world away in East Asia. Today, we count some 85 plant genera known only in these widely separated places. History, rather than contemporary environments, was a critical factor in explaining this distribution.

EFFECTS OF CLIMATIC SHIFTS

In the Cretaceous Period (66–145 Mya), North America and Eurasia formed a supercontinent, Laurasia. Climates were warmer and temperate trees were located at high latitudes, from North America to Europe and Asia. When climates cooled, temperate trees moved southward. As the interior of continents became drier, these trees survived best where summer rainfall persisted, thus in eastern North America, northern Europe, and East Asia. Finally, repeat cycles

---

CONTINENTAL DRIFT AND BIOGEOGRAPHY

Prior to the confirmation of continental drift in the 1960s, some tree distributions were viewed as enigmatic. Some genera (like *Magnolia*) are only native to eastern North America and East Asia. This pattern was called Gray's Puzzle, after botanist Asa Gray (1810–1888). Other genera (like the southern beech / *Nothofagus*) are only found in South America and Australasia. Now we regard Gray's Puzzle as a reflection of the supercontinent Laurasia (now North America and Eurasia) and *Nothofagus* as a reflection of the supercontinent Gondwanaland (now South America, Africa, and Australasia).

of glaciation caused extinctions. Europe suffered the most and East Asia the least. Some temperate tree genera, like *Magnolia*, only survived in eastern North America and East Asia; some, like beech (*Fagus*), survived in both areas as well as Europe; and some, like maple (*Acer*), have wider distributions. Other

↑ Striking fall color in a Japanese deciduous forest. This rich array of autumnal colors reflects the high diversity of tree species growing across this mountainous landscape.

genera survived only in East Asia, including "living fossils" such as the dawn redwood and *Gingko*, which were known as fossils before being discovered in the wild.

The result of these changes is that there is now a diversity anomaly: the three areas of deciduous forest have similar environments but different diversities, with East Asia the highest, and Europe the lowest. The same pattern holds for tropical rainforests (Indo-Pacific, greatest; America, intermediate; and Africa, least) and mangrove forests (Indo-Pacific, greatest; and the Americas, least).

MORPHOLOGY

Leaves come in an array of designs and arrangements. They can be needle-leaved, scale-leaved, or broad-leaved. Evergreen leaves persist for one year or more, creating overlapping sets of leaves from year to year. They are thicker and tougher than deciduous leaves, which are produced for a single growing season.

Leaf arrangements can be alternate (staggered), opposite (paired), or whorled. Leaves can be undivided (simple) or divided into leaflets (compound). If leaflets are further divided into smaller divisions, they are described as twice (or even more) compound. Simple leaves can be lobed or unlobed. Leaves can be smooth-edged (entire) or toothed (with jagged edges). In this chapter, we'll discover that the frequency of these characteristics changes across climate zones, indicating that leaf morphology is part of the adaptive strategy of trees.

We also describe how leaves underlie the flow of energy in trees and forests, being a critical intersection of water and carbon fluxes, and thus important to our understanding of the global effects of warming climates and increasing levels of carbon dioxide in the atmosphere.

← Although most needle-leaved trees are evergreen, larches are deciduous, taking on yellow to golden hues before falling from the tree.

↙ *Sassafras* is one of the few tree genera to produce variable leaf shapes on a single branch.

→ Leaves can be (A) needle-leaved, (B) scale-leaved, or (C) broad-leaved; (D) evergreen or (E) deciduous; (F) alternate, (G) opposite, or (H) whorled; (I) simple, (J) pinnately compound, or (K) palmately compound; (L) unlobed or (M) lobed and (N) untoothed (entire) or (O) toothed. In doubly compound leaves, the leaflets can themselves be divided (not shown, but see page 102).

LEAF CHARACTERISTICS

L eaves can vary even on a single tree. For instance, "sun-leaves" are often smaller, more deeply lobed, and thicker than "shade-leaves." A few genera, like mulberries (*Morus*) and *Sassafras*, produce variable shapes, regardless of light, on the same tree. Famously, eucalypt trees show marked difference between juvenile and adult leaves. However, variations within a species or genus are usually less pronounced than those across many species.

THE EFFECT OF CLIMATIC DIFFERENCES

Beginning in the 18th century, naturalists began to notice that the size and shape of tree leaves varied across climate zones. By the early 1900s, paleobotanists were even using the characteristics of fossil leaves to reconstruct the climates of the past. A recent analysis by a team led by Dr. Daniel Peppe, of Baylor University, found that leaf morphology varied from warm to cold climates and from wet to dry climates. From warmer to colder climates, the percentage of trees with toothed leaves increased over fifteenfold, the number of leaf teeth increased twofold (or more), and leaves were more likely to be lobed or divided. On a gradient from wetter to drier climates, leaves became

---

ENVIRONMENTAL LEAF MORPHOLOGY

Convergent leaf morphology is frequent but not universal since trees can also evolve different strategies for the same set of environmental conditions. For instance, some deciduous species in Mediterranean climates are leafy only in rainy seasons and the leaf blades can be unlike those of their evergreen sclerophyllous neighbors.

Evolutionary divergence across environments sometimes accompanies convergence within one environmental setting. For example, magnolias, oaks, and hollies have deciduous species in cooler environments and evergreen species in warmer ones.

about twentyfold smaller and teeth were small or absent. In tropical climates, leaves are generally larger and evergreen, compared to seasonal, colder, or drier climates. Trees in the Southern Hemisphere showed the same trends as those in the Northern Hemisphere, but, for a given climate, had only about half the percentage of toothed species compared to Northern Hemisphere trees. One explanation is that Southern Hemisphere forests have a higher percentage of evergreens (Peppe's team also found that, in general, broad-leaved evergreens were less likely to have marginal teeth than deciduous species).

We can't always explain morphological trends, but one hypothesis is that leaf teeth and lobes enhance early season photosynthesis in deciduous trees. Drs Kathleen Baker-Brosh and Robert Peet created an early test that supported this idea in 1997. Toothed or lobed margins also allowed leaves to expand more quickly in the spring and early spring leaves had proportionally larger teeth than older leaves.

CONVERGENT AND DIVERGENT EVOLUTION

Trees from different lineages can evolve similar leaf characteristics, a phenomenon called convergent evolution. For instance, trees from a wide array of families have large, evergreen leaves in wet tropical forests. Evergreens have also adapted to nutrient-poor soils and their leaves hold on to nutrients for longer—nutrient retention compensates for the scarcity of nutrients. Two other examples of convergence are:

- Sclerophyllous trees: In summer-dry "Mediterranean" and fire-prone climates, species with small, undivided, hard, evergreen leaves are usual. Leaves with this morphology are called sclerophyllous, referring to the sclerenchyma tissue providing the hard structures of plants.

↓ Drip tips occur in many tropical trees and help the leaves shed water, thereby preventing overgrowths of algae or fungi.

- Deciduousness: The shedding of leaves at the beginning of dry or cold seasons, along with the flushing of new leaves when the seasonally unfavorable conditions have ended, has evolved independently in multiple tree lineages.

TYPES OF TREES

Trees can be classified in multiple ways. In this book, we look at their evolutionary relationships (see page 12), architectural forms (see pages 58–69), life histories (see page 94), and, in this chapter, leaf characteristics.

We start with three basic leaf types: needle-leaved, scale-leaved, and broad-leaved. This classification overlaps with that based on evolutionary relationships, since most needle- and scale-leaved trees are gymnosperms and most broad-leaved trees are angiosperms (*Ginkgo biloba*, a broad-leaved gymnosperm, is an exception). While scale-leaved trees are always evergreen, the other two types can be evergreen or deciduous. This gives us two groups and five basic tree types.

↓ *Ginkgo biloba,* native only to China but widely used in the temperate zone as a city street tree, is a broad-leaved, deciduous gymnosperm with brilliant fall color.

Needle- and scale-leaved trees have strong apical dominance and relatively narrow, spire-like crowns:

DECIDUOUS LEAVES

The seasonal shedding of leaves by deciduous trees at the start of unfavorable times, whether cold or dry seasons, is not a passive response, but a genetically programmed process directed by the tree itself. Trees form a weak zone of cells at the leaf base called the abscission layer, which both seals the stem and leads to leaf fall. Remarkably, this process also shows that leaves can measure time relative to the 24-hour clock, evaluate the number of hours of daylight and dark, and detect temperature trends. Trees use these cues to predict when the benefits from continued photosynthesis are less than the risk of premature leaf death.

- Needle-leaved, evergreen trees include pines, spruces, firs, redwoods, hemlocks, yews, and cedar of Lebanon of the Northern Hemisphere, the Norfolk Island pines and monkey puzzle of the Southern Hemisphere, and *Podocarpus* conifers of tropical and subtropical areas.
- Needle-leaved, deciduous trees include larches, bald cypress, and dawn redwood.
- Evergreen trees with scale-like leaves include cypresses, junipers, and cedars.

Broad-leaved trees have broader crowns and rounded profiles:

- Broad-leaved, evergreen trees include rhododendrons, eucalyptus, and some southern beeches, as well as figs and thousands of other tropical trees.
- Broad-leaved, deciduous trees include maples, beeches, ashes, oaks, chestnuts, alders, birches, willows, aspens, cottonwoods, elms, walnuts, plane trees, lindens, locusts, buckeyes, acacias, and some southern beeches.

Several other groups with tree-sized species fall outside this classification, possessing traits that are distinct in their own right: the palms, tree ferns, and succulents.

PHOTOSYNTHESIS

In 1634, Belgian scientist Jan van Helmont initiated a project, often considered the first biological experiment, to test the then-common idea that plants grew by assimilating soil. He weighed a willow stem, planted it in potted soil that he had also weighed, and added water when needed. After five years, he reweighed the willow and soil. The willow had gained 164 lb (74 kg), while the soil had lost only 2 oz (57 g)! Van Helmont rejected the idea that the soil provided for the weight gain. However, he concluded (wrongly) that the weight gain had come from water.

In retrospect, the experiment had several flaws (for example, van Helmont did not weigh the water inputs nor account for evaporation). He also failed to consider the possibility of an atmospheric source for the weight gain: carbon dioxide. Van Helmont's work paved the way, but the discovery of photosynthesis, the process by which plants use sunlight, carbon dioxide, and water to produce energy-rich compounds (in the form of sugars) and oxygen had to await the work of Jan Ingenhousz in 1779.

↙ In willows (*Salix*), male catkins (upper left) are produced on different individuals than female flowers (lower right). Willow leaves (lower left) are typically long and narrow.

→ (A) Sunlight supplies energy for photosynthesis in which (B) carbon dioxide (CO_2) is absorbed from the atmosphere and (C) minerals and (D) water (H_2O) are taken up from the soil. The source of soil moisture is precipitation. (F) oxygen (O_2) is released as a by-product. Water vapor leaves the tree through (E) evapotranspiration.

$$\text{Sunlight}$$
$$6CO_2 + 6H_2O \longrightarrow 6O_2 + C_6H_{12}O_6$$
$$\text{Chlorophyll}$$

41

PHOTOSYNTHESIS, CO$_2$, AND H$_2$O

Photosynthesis transforms carbon dioxide (hereafter CO$_2$) and water to sugars and oxygen. The sugars provide the energy source for plant growth and reproduction. Photosynthesis takes place in chloroplasts, organelles that contain the energy-capturing compounds, chlorophyll a, chlorophyll b, and accessory pigments (carotenoids). Leaves are green because chlorophylls use less of, and therefore reflect, the green part of the solar spectrum.

Water and inorganic nutrients from the soil are transported to the leaves via the thin tubes of the xylem, one of two types of vascular tissue in higher plants (the other is phloem). The evaporation of water from leaf surfaces, called "transpiration," creates tension within the

---

THE ROLE OF STOMATA

To regulate water and CO$_2$, leaves have surface pores called stomata. These are bordered by guard cells that can widen or shrink the size of the opening. The stomata are an important crossroads, with water vapor leaving and CO$_2$ entering the leaf through these pores. This creates a potential conflict, however: opening the stomata to absorb CO$_2$ necessarily increases water loss, while closing the stomata to conserve water decreases CO$_2$ uptake.

→ Stomata are leaf pores that serve as the crossroads of water vapor loss and carbon dioxide gain by leaves.

xylem, pulling water upward. Maintaining water flow is critical; severe droughts can cause a permanent interruption of the water column when bubbles of water vapor form (known as "cavitation").

THE BOUNDARY LAYER

Leaf characteristic can also affect the thin layer of air (the boundary layer) next to the leaf surface. This layer determines the rate of movement of CO_2 and water vapor close to the surface, as the concentrations of these substances are different within the boundary layer compared with the external atmosphere. While the thickness of the boundary layer is influenced by temperature and wind, it can also be affected by leaf morphology. Leaves that are undivided or unlobed have thicker boundary layers. Some plants have hairy leaf surfaces, which increases the thickness of the boundary layer. Boundary layers can be more humid than the external environment, thereby slowing water vapor loss from the leaves.

THE ROLE OF PHLOEM AND XYLEM

Sugars produced in the leaves are transported throughout the tree in the vascular tissue. Annual increments of phloem and xylem are produced in the tree stem by the cambium, a ring of undifferentiated tissue. The phloem is produced to the outward side of the cambium and the xylem to the inward side. The increasing diameter of the tree trunk reflects the concentric rings of xylem that are produced—the phloem is regenerated, but older cells do not persist. The xylem becomes a non-living tissue with only a one-directional (upward) movement of water, which is governed by physical forces (the tension created by transpiration). The phloem is a living tissue that moves sugars, amino acids, and other organic substances from the leaves to other parts of the tree.

While vascular tissue is hidden in the trunk, it is likely that all of us have had at least one direct experience of its function. Many trees in seasonal climates mobilize sugars rapidly at the beginning of the growing season to fuel the production of new leaves. This "sap" run in sugar maple trees is collected and boiled, becoming maple syrup in the process. Indigenous populations also made sugary syrups from other trees.

PHOTOSYNTHESIS AND CLIMATE CHANGE

Climate warming impacts natural systems in diverse ways, from rising sea levels to catastrophic fires and extreme weather events such as floods and droughts. Here, we address only the effects on photosynthesis, which occur through two interconnected pathways:

Firstly, warming increases the rates of transpiration and respiration. Closing stomata can conserve water, but this restricts CO_2 uptake, potentially reducing productivity. Warming increases leaf respiration, resulting in the release of CO_2. There is also a possible positive effect: warming can result in longer growing seasons (unless other factors become limiting). Warming effects depend on water supply, though future changes in precipitation are harder to predict than temperature changes. Secondly, CO_2 is a raw material for photosynthesis, so added CO_2 increases productivity (again, unless other factors become limiting) and carbon storage.

PHENOLOGY

Phenology is the study of the timing of biological events such as flowering, fruiting, and leafing. Trees sense stimuli such as light, temperature, and the time of day and, in essence, predict when favorable conditions may occur. Thus, an area of climate research is whether events like flowering are occurring earlier in the year. Beyond shifts in timing, however, a deeper concern is that a tree's calibration of the relationship between day length (a function of latitude and season) and temperature (which has considerable variation) will become incorrect and their responses non-optimal. Mismatches between the responses of trees and their pollinators or seed dispersers may also occur if organisms shift in different ways due to climate warming.

The balance of negative and positive effects on photosynthesis varies geographically and over time. A study by Martin Venturas of the University of Utah and colleagues concluded that the amount of warming and CO_2 increase was less important than the ratio between them. In other words, if temperature rises faster than CO_2, respiration losses will outpace photosynthetic gains. It is concerning that respiration increases with temperature, but photosynthesis reaches a plateau, suggesting warming effects will become increasingly negative.

The leaves themselves can play a role in how much climate change affects photosynthesis through acclimation, in this case a shift in size and shape, as well as the density of leaf stomata. At present, this is one of the uncertainties in predicting the effects on photosynthesis.

← Towers such as this one in a European beech forest monitor concentrations of CO_2 and other greenhouse gases to gauge effects on tree productivity. In some experiments, towers release CO_2 to test effects of future increases.

GLOBAL PATTERNS OF HEIGHT

In 2010, Professor Michael Lefsky, of Colorado State University, analyzed satellite-based measurements from an instrument originally designed to measure the growth and melting of glaciers by observing changes in the height of the Earth's surface. The measurements were taken using a laser from the satellite bounced from the Earth's surface, through the forest canopy, and back up to the satellite. Lefsky cleverly realized that this data could also provide a space-based reconnaissance of the heights of the world's trees.

Forests composed primarily of tall trees were found in hot, moist locations with plenty of sunlight—the tropical wet forests of South America, Africa, Indomalaya, and New Guinea. Tall forests were also found in the Pacific Coastal region of Canada and the United States and the Appalachian Mountains of the eastern United States. Forest clearing has reduced the presence of tall trees in agricultural zones in Europe, North America, and Asia except for a few remnant forests, which are mostly in mountainous terrain.

← Large trees such as this red oak (*Quercus rubra*) expand at the base, a phenomenon known as "butt-swell." In the USA, a tree's diameter is taken at a height of 4½ ft (1.37 m) to avoid errors due to butt-swell.

← The tallest wooden structure is Poland's Gleiwitz Radio Tower. Completed in 1934, and built with Siberian larch (*Larix sibirica*) with brass connectors, it stands 364 ft (111 m) tall.

→ Shown here in Borneo, dipterocarp trees (in the family Dipterocarpaceae) are found across the tropics in tropical lowland rainforests. Dipterocarp trees in Southeast Asia are among the tallest trees on Earth. They emerge from the subcanopy and tower over the other trees.

MECHANICS OF
GROWING TALL TREES

At the time of the American Revolutionary War, a first-rate English ship of the line had a mainmast with a base diameter of 40 in (1 m). The mast height of a tall ship was a yard of mast length per inch of diameter. Ship masts have a perverse tendency to break at the most inopportune times—in battle or during heavy storms. The danger of installing massive masts on a rolling ship in wind and heaving waves created a keen interest in engineering light but strong ship's timbers.

WITHSTANDING STRUCTURAL DAMAGE

In 1973, the late Professor Thomas McMahon, of Harvard University, argued on mechanical grounds that a tall, self-supporting, tapering column (like a tree trunk) should have a height proportional to its diameter raised to the power of $2/3$. When one looks up from the base of a large tree, the taper of the trunk implied by this relationship is

WHAT RESTRICTS TREE HEIGHT?

Beyond the mechanical limits of the heights of wooden structures, Professor F. Ian Woodward (formerly of Sheffield University) noted that trees are limited in height by four physiological constraints:

1. There is a limit on how high transpiration can lift a column of water in the transporting xylem tissues against gravity and friction before undergoing embolism and the formation of air bubbles.

2. The upward flow of water decreases with height, which decreases growth capacity.

3. The increase in investment in non-photosynthetic, structural leaf tissue with height decreases photosynthetic efficiency.

4. The rate of diffusion of CO_2 needed for photosynthesis into the leaf decreases with height.

These constraints imply a maximum tree height of 400–427 ft (122–130 m).

obvious. When McMahon considered over 700 record trees of different species, he found that trunk heights and diameters were typically about four times stronger than necessary to prevent buckling under their own weight. This "over-engineering" is to be expected. A trunk should have additional strength beyond that needed to stand under its weight if the tree is to survive stronger winds. In forests, tree saplings are often bent or buckled, implying that they are closer to the edge of buckling. One also sees buckled trunks and other damage in forests after strong winds, hurricanes, or tornados.

← General Sherman is a giant sequoia located 6,919 ft (2,109 m) above sea level in the Giant Forest of Sequoia National Park in Tulare County, California. By volume, it is the largest known living single-stem tree on Earth. It is thought to be around 2,200 to 2,700 years old.

TRUNK STRENGTH TRADE-OFFS

In 2009, Professor Jérôme Chave and colleagues, of the University of Toulouse, examined the functional attributes of the woody tissue of 8,412 tree species, including density, biochemistry, anatomy, and the mechanical strength of their wood.

WOOD DENSITY

Wood density is determined as the weight of wood relative to wood volume after prolonged drying in an oven. Wood density is positively related to the amount of stress needed to break wood, the elasticity of wood, the resistance to splitting along wood fibers, and the work needed to break a piece of wood. Wood density across thousands of tree species worldwide accounts for 43 to 77 percent of the variation in these mechanical features. The density of wood indexes the allocation of a tree's photosynthetic production to trunk volume and suggests "economic" trade-offs in the growth of trunks. Is it a better strategy to allocate more photosynthate to building strong, elastic, hard-to-break trunks, or to invest less in dense wood but grow taller and faster and capture a greater portion of sunlight?

WOOD DENSITY VARIATION

Wood density varies within individual trees. As a tree grows, for both the trunks of conifers and angiosperms, interior sapwood layers are converted into heartwood by filling water conduits and pores with polymerized organic biochemicals. As a result, heartwood densities are significantly higher than sapwood densities. The proportion of heartwood to softwood in a tree is larger at the base and grows systematically smaller with height. For this reason, larger trees have proportionally higher wood densities than smaller trees of the same species.

VARYING WATER FLOW VELOCITIES

The water conduits in ring-porous angiosperms are arranged in rings, often annual rings. They are scattered in diffuse-porous angiosperms. These water transport systems differ in the speed that the sapwood can transport water. Maximum flow velocities range from 0.3 to 0.8 mm per second in conifers, 0.2 to 1.7 mm per second in ring-porous hardwoods (poplars, maples), and 1.1 to 12.1 mm per second in ring-porous hardwoods (ash, elms).

WATER HIGHWAYS

Not only do trunks require mechanical strength, but they must also conduct water from the tree's roots to the leaves. The use of water by the leaves is generated by intermeshed needs:

1 To supply the tree with water (H_2O) for photosynthesis. When energized by the energy of incoming light, the photosynthesis process combines carbon dioxide (CO_2) and water to produce sugars and oxygen (O_2).
2 To supply the water that evaporates from and cools the leaves, bringing leaf temperatures into a livable range.
3 To provide water to prevent leaf wilting.

In both conifers and angiosperms, the sapwood conducts water. In conifers, sapwood xylem cells are hollow, tubelike structures called tracheids. Thin-walled tracheids move water and thick-walled tracheids supply mechanical strength. In angiosperms, the sapwood xylem cells are differentiated into structures called pit vessels, which are specialized to move water, and wood fibers that provide strength.

Higher flow speeds mean different trees need less sapwood "plumbing" to move water upward. In ring-porous angiosperms, water may move through sapwood rings laid down over the past one to three years. This allows them to rapidly reconfigure their water transport systems when conditions change.

APICAL DOMINANCE

As a tree grows from a seedling, the main stem develops more strongly than the side stems. This phenomenon is known as apical dominance and is attributed to the effect of a plant hormone called auxin, which, when it is produced by the apex bud of a stem, inhibits the outgrowth of other auxiliary buds.

Auxin, produced by the main bud, moves down the shoot and suppresses the outgrowth of auxiliary buds. The high demand for sugars by the growing tip could augment this suppression by heavily using plant sugars and suppressing lateral growth. Trimming or herbivores grazing on central buds removes this apical control and can result in bushier plants. One result of this is that small tree seedlings often have a pleasing, crystalline regularity. This regularity is also seen at the tops of trees where the apical dominance allows for tree-determined geometry.

↓ Horticultural varieties of tree variation in apical dominance were used in landscaping, as seen in the apically dominant black poplar (*Populus nigra*) alleys in France.

↘ Horticultural cultivars that lack apical dominance take on a spreading form, as seen in this Sargent's weeping hemlock (*Tsuga canadensis* 'Pendula').

→ Engelmann spruce (*Picea engelmannii*) as well as other spruces (*Picea*) show strong apical dominance and a resultant conical form. This form makes the tree more efficient at capturing horizontal incoming light.

HOW TO BE A HYPER-TALL TREE

The degree of apical dominance influences the ratio of tree height versus the width of the tree's crown. We discuss how trees develop extraordinary widths on page 57. Here, we consider how two of the tallest trees, the mountain ash (*Eucalyptus regnans*) and coast redwood (*Sequoia sempervirens*), resemble each other ecologically.

The world's tallest tree is a coast redwood called Hyperion, which stands at 380.3 ft (115.92 m), and the second tallest is a mountain ash named Centurion, which is 329.7 ft (100.5 m) tall. The redwood is a conifer and the eucalypt an angiosperm. Although separated by thousands of miles, they grow in relatively similar climates, according to Professor Elgene Box of the University of North Carolina: "both occurring in the transition from Mediterranean to marine west-coast Climate." These winter-wet, summer-dry, and often foggy forests can suffer horrific wildfires with devastating fire-tornadoes and other firestorm effects.

WILDFIRES: A TALE OF TWO TREES

Mountain ash has a restricted range and is found in the Australian states of Victoria and Tasmania. It is a valuable timber tree used for furniture construction, trims, and general construction. A member of the *Eucalyptus* sub-genus *Monocalyptus*, like other monocalpyts (and unlike most other eucalyptus species), mountain ash does not sprout when damaged by wildfires. Instead, its regeneration requires large wildfires so hot that they leave a sterile soil covered with a fine white ash.

Coast redwoods are restricted to small areas in California and are the state's most valuable timber species. The trees have thick bark, an adaptation to help them resist wildfires. The species can be regenerated from seeds or cuttings from saplings and has been successfully used in forest reclamation projects. The trees are considered resilient to wildfire and can sprout after fires from the roots and dormant buds in the trunks.

These very tall trees differ in several ways. The coast redwood is one of a species with 1,000-year-plus ages. It can resist fires and is, in general, adapted to regenerate well after a wildfire. The mountain ash is shorter lived (only 400 years). It is sensitive to the powerful wildfires typical of its habitat and requires recovery time to recruit new seed trees before the arrival of a second fire. Over geological time scales, *Sequoia* was widespread and successful. Hopefully, it will remain so, even if the fire climate is changing. The same should be said of *Eucalyptus regnans*.

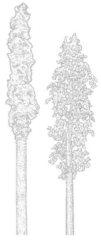

→ Hyperion, the tallest tree, is a coast redwood from California, and Centurion is a *Eucalyptus regnans* from Australia. Both are found in similar climates with wet winters and dry, fire-prone summers. But they differ in their ecological adaptations to similar climatic settings.

Hyperion   Centurion

MOUNTAIN ASH FIRE VULNERABILITY

Wildfires hot enough to provide the sterile ash beds and an optimal seeding site also destroy mountain ash parent trees. It then takes years for mature, seed-producing trees to grow. A young mountain ash forest may regenerate and grow from the ashes of hot wildfires, but it takes 20 to 30 years before any of this generation of trees becomes large enough to resist light surface fires. The fire-generated new forest is thus fire-vulnerable during this time. When fires are too frequent, the forests burn a second time before the young trees can produce seed, and this too-rapid return of wildfire eliminates mountain ash from the site. Since only hot fire produces the necessary sterile soil, if wildfire does not burn the site in a tree's lifetime, then mountain ash is lost from the site. These unburned forests are likely to become temperate rainforest.

NON-MONOPODIAL
LARGE "TREES"

The biggest trees are not the tallest. Instead, they reach their immense mass in two ways: either dropping trunks down from their canopies or growing trunks up from their roots. Both approaches produce prodigiously large trees, assuming your definition of a "tree" includes multi-stemmed plants.

BANYANS AND CLONES

The "down from the canopy" method is common in figs (*Ficus*) like the banyan (*Ficus benghalensis*), the national tree of India. A striking specimen is the Great Banyan in the Acharya Jagadish Chandra Bose Indian Botanic Garden in Howrah, India, which has 2,880 prop roots and covers 3.71 acres (1.5 hectares).

ROOT-GRAFTED TREES

Below ground, roots from a single tree can join or graft together—as can those from different trees of the same species—to form connected trees, which, like clones, have an extended shared root system. This can also occur among trees of different species. Usually, the less related the species of individual trees, the less likely their roots are to graft. Root-grafted trees are able to share resources, which can mutually promote the overall vigor of a stand of adjacent trees, but there are potential downsides to this mutual sharing of root resources. For example, plant diseases can spread rapidly across shared root systems. In a bizarre case from New Zealand, a single, large, living stump of an ancient kauri (*Agathis australis*) tree survives as a "vampire tree"—drawing water and nutrition from root systems shared with its neighbors.

Clonal trees, in contrast, sprout what will become trunks from an extended root system. Pando is the name of a clonal colony of an individual male quaking aspen (*Populus tremuloides*) growing from a common rootstock in the Fishlake National Forest, in Utah, in the United States. It is a gigantic, single, individual organism with identical genetic markers throughout. It occupies 108 acres (43.7 hectares) and weighs an estimated 6,600 US tons (6,000,000 kg). Pando is a consequence of the ability of quaking aspen, as well as other tree species, to throw out lateral roots that then send up erect stems. Repeat the process for several thousand years and one ends up with a super-clone such as Pando. Pando is thought to be over 10,000 years old.

← Quaking aspens sprout from their roots to produce multiple trunks over large areas. Some of these clonal trees are among the Earth's largest organisms, living for thousands of years.

CROWN SHYNESS

When one looks straight up through a forest canopy, the tree crowns typically do not touch, a phenomenon called "crown shyness." This is emphasized in images taken with a "fish-eye" camera lens, which allows views of canopies as if they were displayed on the inside of a hemisphere.

There are multiple explanations of crown shyness. For example, winds can whip twigs on the branches of adjoining trees together, knocking off buds and blocking the growth of the crowns into each other. This "crown-pruning" mechanism also affects tree spacing in forest plantations. Crown shyness can also arise as a result of shade avoidance in plant growth. A proportional increase in far-red light indicates the presence of a nearby plant and plants tend to grow away from elevated far-red light. Plants also grow away from increased blue light to avoid shade. Crown shyness among trees in a forest hint at a mechanism for ecosystem self-organization—a theme we will explore further in this chapter.

↓ Crown shyness in a *Dryobalanops aromatica* tree canopy, Forest Research Institute of Malaysia (FRIM), Kepong, Malaysia. Note the spaces outlining individual tree canopies.

→ If one walks through a relatively dense forest of trees of similar heights and sizes, the regularity of the canopy crown shyness can be striking. Once one learns to look for such geometry, it becomes clear that this pattern is widespread.

HALLÉ/OLDEMAN MODELS

Tropical ecologists Francis Hallé and Roelof A. A. Oldeman, subsequently joined by Philip B. Tomlinson, proposed the concept of architectural tree models based on the production, activation, and growth of meristematic tissues. Meristematic cells, which can be programmed to develop into flowers, shoots, leaves, or bark, and so on, reside in buds (protective protuberances, or "bumps," on the stems of vascular plants) and elsewhere in the tree. Buds can be triggered to grow or can be suppressed by biochemical signals from other tree tissues. These interactions work in concert to selectively enhance or suppress the growth of meristematic tissues from other buds in tree crowns. This programming can produce almost crystalline forest geometries.

DIFFERENT TREE MODELS

For the trees themselves, other simple "rules" involving temporal changes in buds can produce an astonishing range of tree shapes. In the Hallé/Oldeman models, the timing and patterns of bud-break and flower production, as well as the rate and patterns of stem growth, produce a tree's geometry. There are 23 such architectural models, each named after a prominent botanist.

The simplest of the Hallé/Oldeman models is Corner's model, in which the meristem tissue in a bud at the top of a single, unbranched trunk grows to trunk height and then the trunk produces a flower. Afterward, the entire plant dies and seeds produced by the flower repeat the cycle. In the more complex Nozeran's model, a bud at the apex of the plant produces a tier of horizontally oriented branches; a new vertical stem then arises from below this tier. The bud at the top of this new stem repeats the process to form additional horizontal tiers.

↑ Found across the
eastern United States,
staghorn sumac
(*Rhus typhina*) grows
according to

Leeuwenberg's model.
Twigs will grow from
a pair of buds below
each seed cluster the
following spring.

LEEUWENBERG'S MODEL

Leeuwenberg's model is slightly more complex. Here, the tree grows
in modules. A primary shoot called a sympodium lengthens to
produce a growth unit. The apical meristem at the end of this shoot
dies, often because it is irreversibly programmed to produce flowers
or leaves. Further growth is then initiated, often from two activated
lateral buds. The tree is formed of a collection of "Y-shaped" units.
The stem elongation, flowering, and activation of growth from the
lateral meristems often occur simultaneously and in sequence in
the growing tree.

CROWN GEOMETRY: MONO- AND MULTI-LAYERED TREES

There are trade-offs in a tree's ability to survive under relatively low light conditions versus high light conditions, which consistently emerge at multiple scales from leaf to forest canopy. These trade-offs are sometimes posed as economic questions: How much of an essential nutrient (or of photosynthetically produced sugars or of water) should be invested in growing more leaves? . . . in growing taller trunks? . . . in developing more roots and root surfaces? . . . in making more seeds?

SUN-LEAVES AND SHADE-LEAVES

At the level of a single leaf, one can see similar patterns in trade-offs operating at different levels through the canopy. The leaves in the top of a tree's canopy are smaller but thicker because they contain more cells with chloroplasts full of chlorophyll. They are richer in nitrogen and other nutrients. If the leaves have lobes, they are more deeply incised. In the fall, look at fallen leaves beneath a deciduous tree. The "sun-leaves" from the top of a tree are strikingly different from the "shade-leaves" from the lower canopy. So much so that the leaves seem as though they could be from different species.

MONO- AND MULTI-LAYERED TREES

The same trade-offs are seen in sun-leaf/shade-leaf morphologies. Imagine the differences between sun-leaves and shade-leaves did not exist and that a tree had the same kinds of leaves. What might be the best way to arrange these leaves? The late Princeton Professor Henry Horn considered this question and investigated the theoretical consequences of two extreme cases: mono- and multi-layered trees.

At high light levels, a multi-layered tree can capture the available incoming light and undergo maximum photosynthesis. Under lower light levels, the amount of light absorbed to drive photosynthesis is reduced and some of the shade-leaves become net losers. Mono-layered canopies underperform multi-layered canopies when there

FUNCTIONAL DIFFERENCES

Differences in leaf appearance between sun-leaves and shade-leaves are paralleled by differences in function. In low light, shade-leaves have more net photosynthesis—because they bear lower "costs" from spending photosynthetic capital on maintaining non-productive tissue. Sun-leaves in low light use more energy to maintain their capacity for high rates of photosynthesis, a capacity that cannot be used in lower light. At high light levels, shade-leaves cannot take advantage of the extra light. Relative to shade-leaves, sun-leaves have higher productivity in full light but lower, even negative productivity in low light.

is ample light. At low light levels, mono-layered canopies can survive a shade so deep that multi-layered canopies are in a negative balance and perish.

Multi-layered trees are likely to survive in young forests or those that have been destroyed by fire, clearing, and similar events. In this scenario, rapid growth with high rates of photosynthesis would favor the multi-layered tree. But mono-layered trees are destined to play a long game—surviving on the shady forest floor, waiting for an overhead tree to die, and taking opportunities to grow a little larger.

→ The bunya pine (*Araucaria bidwillii*) is a native tree in Queensland (Australia) rainforests. The tree has an open but functionally multi-layered canopy, meaning that a beam of light from any point above strikes upon more than one group of leaves.

ADAPTIVE GEOMETRY
OF TREE CROWNS

The overall geometry of tree canopies changes with the environment. In the northern boreal forests of Eurasia and North America, spruces (*Picea*) and firs (*Abies*) show a remarkable capacity to capture light due to the arrangements of their leaves and are often referred to as "dark conifers" for this reason.

Black spruce (*Picea mariana*), for example, which covers vast areas of Canada and Alaska, has light-capturing arrangements of needles. As NASA's Dr. Kenneth Jon Ranson explains, this is to ensure that "photons go in but they don't come out." There is clearly a good reason for the common name, black spruce. At the whole tree level, black spruce presents as a tall, narrow spire of needles that go all the way down to the ground. They capture the flat sunlight that is associated with the near-horizon location of the sun at high latitudes.

↓ The multi-angled orientations of black spruce (*Picea mariana*) leaves are very effective at capturing light from different angles.

↓ Vertical leaves of alpine ash (*Eucalyptus delegatensis*) capture morning and sunset light, when conditions in the Australian Alps are more favorable for photosynthesis.

→ The vertical canopies of black spruce trees serve to increase the capture of direct sunlight found under the light conditions in the high-altitude boreal and Arctic regions. A boreal tree can shade or be shaded by a surrounding tree, depending on the time of day.

SHINOZAKI
PIPE MODEL THEORY

One of Leonardo da Vinci's most famous drawings is his *Vitruvian Man*, which depicts a nude male figure with two positions for the arms and legs, one inscribed on a circle and the other on a square. The drawing depicts the "perfect" proportions of the human body as described by the Roman architect Vitruvius between 30 and 15 BC.

TREE PROPORTIONALITY

A less well-known drawing by Leonardo depicts regular proportions in tree branching. His mathematical rule for tree proportionality was that the summation of the cross-sectional area of all the tree branches above a branching point at any height equals the cross-sectional area of the trunk or branch immediately below that branching point. Puzzling over and testing this rule—"Is it correct? How is it is caused?"—has continued to the present day.

For example, Kichiro Shinozaki and an interdisciplinary group of colleagues (K. Yoda, K. Hozumi, and T. Kira) at Osaka City University, in Japan, launched investigations in 1964 that have continued for over 50 years in pursuit of a basis for Leonardo's rule. Shinozaki's Pipe Model represented a tree as a collection of "pipes" carrying water from the roots to the leafy canopy. The resultant plumbing would have a constant cross-sectional area of sapwood from the bottom to the top of the tree. With the pipe model, if the cross-sectional areas of pipes in a tree implies the mass of leaves in a tree, then the number of pipes in a forest would imply the mass of leaves in the forest. The expectation is that the cross-sectional area of sapwood summed over an area of a forest should have a strong relation with the mass (or area) of leaves over the same area.

→ An aerial view of rainforest canopy in Bellenden Ker, North Queensland, Australia. Photograph taken in November, 1989.

FOREST REGULARITY

The connection between leaves and sapwood cross section encoded in the Shinozaki Pipe Model, and the leap to enlarge this to the forest scale, implies rules for a constancy in the formation of a forest's geometry. There are many regularities in forests. For forest plantations and in natural forest ecosystems, the cross-sectional area of "pipes" supplying water to leaves show consistent ratios, as do other measurements of tree geometry.

FOREST PLANTATIONS

O ne of Shinozaki's colleagues, Kyoji Yoda, developed a mathematical model of a phenomenon called self-thinning, the tendency for the logarithm of the numbers of trees comprising a forest versus the logarithm of the average size of the trees in the forest to lie on a line with a slope of 1.5 (an "allometric function").

Called Yoda's self-thinning law, this is another of the interwoven regularities in forests. It is a "packing rule" involving how many trees of a similar size and shape can be packed into a box of a given size. All sorts of natural systems roughly conform to such allometric packing, with the number of things adjusted to fit the size of container. For example: How many corals can grow in a reef? How many oysters can be found in an oyster bed? How many radishes can grow in a pot? And so on . . .

MANAGED FORESTS AND THE SITE INDEX

In practical forestry, there is the analogous case that involves using forest data compilations to determine when to thin (partially harvest) or harvest trees. Modern scientific forestry had its origins in the virtual clearing of the European forests for houses, great wooden ships,

---

SELF-THINNING RULES

Yoda's self-thinning rule has the additional consideration that if the average number of trees predicted for a forest with trees of a given size is too large, then some of the trees would be expected to die, or, in forestry parlance, the forest should thin itself of the extras. Which trees die? Often it is the smaller trees, or those too close to other tree crowns, or tall trees exposed to chilling winds. Whatever the reason, the forest becomes an automaton, thinning itself to agree to a fundamental rule of forest stand geometry as the trees grow larger.

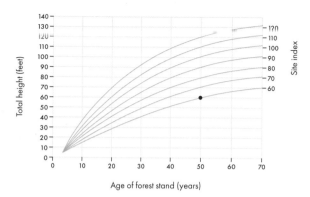

↑ A site index is the
height of trees in a stand
growing in a plantation
at an index age (in this
case, 50 years). From
this, one can predict
future height, growth,
increase in wood, etc.

construction timbers for mines and railway ties, and fuel for houses
and industry. The proper stewardship of natural resources, particularly
forests, was an integrated set of practices called *Nachhaltigkeit* by its
mostly German-speaking developers. Sustainability was first used in
English as a direct translation of *Nachhaltigkeit*. Central to this
practice was the regular compilation of data on the growth, size,
height, and mass of merchantable timber for millions of individual
trees on thousands of forest survey plots. It was found that the tree
height in a forest of a given age (the "site index") allowed the data
to be organized to predict forests over time, which is the preamble to
forestry decisions of when to harvest, when to thin, and so on.

The site index also draws on the quasi-regularity of the descriptors
of forests. These geometric rules may fascinate quantitative ecologists,
as well as provide them with ample opportunities to sharpen their
understandings of the nature of nature. But for many of us, forests are
simply beautiful environments to contemplate—much as crystals, the
rhythm of waves, and the colors of a sunset are beautiful in their
spontaneously generated regularity.

BARK VARIATION

Bark plays a critical role in protecting tree stems from fungi, insects, mammalian herbivores, desiccation, fire, mechanical injury, and temperature fluctuation. Although bark comes in many colors, textures, and designs, it tends to be overlooked in tree identification compared to the leaves, flowers, and fruits. While some species have unmistakable bark, one problem is that the appearance can change with age and diameter. In some trees, there are three or more bark stages: unbroken bark when young, ridged bark in middle age, and smoother and plated bark in old age. Bark is also overlooked because bark traits can be difficult to describe or measure.

TEXTURE AND COLOR

In most trees (dicots and gymnosperms), stems increase in circumference over time through secondary growth. In some species, the bark can remain smooth if its growth keeps pace with this growth (for example, in beech trees). However, for most trees, the stem surface changes from smooth to rough, with the timing of this change varying among species from a few years to a few decades.

The bark surface then takes many forms in different species: vertical ridges with furrows in between, plates, scales, and papery sheets. Some bark is regularly shed and replaced by new layers

---

A PROTECTIVE SHIELD

The outer bark is a mechanical barrier but also a chemical one in that it is low in nutrients and high in lignin, and therefore unattractive to herbivores compared to the living tissues it protects. Bark can also be protected by toxic or repellent compounds, thorns, and warty outgrowths. When injured, bark can exude resins or latex that seal wounds and repel herbivores.

underneath. In "shaggy-barked" species thick plates are partially separated from the trunk. Bark often comes in shades of brown, red-brown, or gray, but it can also be yellow, orange, red, green, or white. The bark of some species consists of multicolored mosaics, as seen in plane trees (see page 72) and rainbow gums (see page 76).

LENTICELS FOR GAS EXCHANGE

While bark forms a protective barrier, it also features "lenticels" that allow for the gas exchange essential to the living tissues beneath the outer bark. Lenticels vary among species, with some forming characteristic horizontal lines and others diamond-shaped or rounded.

↓ Grove of beech trees (*Fagus*) showing smooth, gray bark and darker horizontal lenticels, which allow for gas exchange between the atmosphere and living inner bark.

JIGSAW-PUZZLE BARK

Plane trees (*Platanus*), called sycamores in North America, have remarkable bark, consisting of thin, flat plates that become loose and fall in irregular patches as the tree grows. The bark layers that are revealed vary from yellow to greenish or white, producing a jigsaw-like patchwork. With age, the lower trunk becomes uniformly gray or brown, but the patchwork of colors persists on the upper trunk and branches. This distinctive bark is familiar in temperate cities due to the widespread planting of the London plane (*Platanus × acerifolia*), a hybrid of the North American and Oriental species (*P. occidentalis* and *P. orientalis*, respectively).

Why do plane trees have jigsaw-puzzle bark? Intriguing hypotheses include the following: that caterpillars and adult insects become more vulnerable to bird predation as they cross from darker to lighter bark patches, that the loose bark plates reduce colonization by woody vines, which can potentially shade the crown, that light colors reflect light and reduce overheating, and that bark shedding increases photosynthesis in the revealed bark layers.

↓ The sycamore's jigsaw-like bark develops as the outer plates of bark are shed, revealing lighter-colored plates below.

↓ Sycamores and other plane trees have spherical fruits that are aggregates of small achenes (dry, one-seeded fruits).

→ Wild populations of American sycamore (*Platanus occidentalis*) occur on floodplains and riversides, but this species and hybrid plane trees are widely cultivated. In the pre-industrial landscape sycamore was the largest deciduous tree in eastern North America, reaching 13 ft (4 m) in diameter and 174 ft (53 m) in height.

BARK ANATOMY

Most species of dicot or gymnosperm trees share the concentric arrangement of stem tissues most familiar from the tree rings of the xylem (see page 42). The tree cell types are arranged as follows:

- Pith: At the center of twigs, there is often a narrow cylinder called the pith (this is sometimes hollow, a non-living tissue, and does not change over time).
- Xylem: Next is the xylem, which is formed by annual increments of wood. In seasonal climates, the spring or early xylem is less dense than the summer or late xylem, so the annual growth rings are prominent and create the wood's grain.
- Inner bark: Next comes the "inner bark," comprised of the vascular cambium (also called the lateral or secondary meristem) and phloem. Phloem cells, like xylem, are produced annually but do not form rings; new phloem cells simply replace older ones as the tree grows.
- Outer bark: Beyond the phloem is a protective layer called the periderm or "outer bark." The innermost part of the periderm is the cork cambium (or phellogen), which produces new bark cells. These cells die and become part of the tree's outermost circle, the bark that forms the external "skin" of the tree, called the rhytidome.

IMPORTANT FUNCTIONS

The outer bark protects the living nutrient- and energy-rich tissues of the inner bark. The inner bark and xylem are important not only for the resources they contain, but also because they play three critical roles in tree survival. First, the phloem is the transportation pathway for carbohydrates and other compounds between the leaves and roots. Second, the xylem transports water and nutrients from the soil to the leaves, while also providing mechanical strength. Third, the vascular cambium produces the tree's new xylem and phloem cells.

GIRDLING

Trees with a concentric arrangement of tissues can be killed by "girdling," the severing of the phloem and cambium all the way around the stem, either by cutting just deep enough to reach these layers or through the lethal temperatures that kill stem tissues during fires. After girdling, the trees remain standing and the xylem may continue to supply water to the leaves for a time, but there is no further production of xylem and phloem,

↑ Girdling removes the outer bark, phloem, and cambium, revealing the underlying xylem.

and no transport of sugars from the leaves to the root system. Historically, one way of beginning the clearing of land for agriculture was to girdle the trees before felling them.

TREES WITHOUT CONCENTRIC TISSUES

The concentric arrangement of tissues is not found in monocot trees, such as palms, bamboo, yucca, screw pines, and grass-trees or tree ferns (nor, in fossil examples, trees classified as club mosses or horsetails). Often, the bark in these groups consists of hardened outer cell layers that are the remnants of leaf bases (as in palm trees) or are derived directly from the primary, or apical, meristem.

The lack of a vascular cambium means that stem diameters do not increase, or increase very little, over time, so the bark does not change with age and there are no concentric growth rings. The lack of a cork cambium means there is no mechanism for repair if the bark is injured. In monocots, the phloem and xylem are found in vascular bundles scattered throughout the stem; girdling is therefore ineffective.

BARK VARIATION IN EUCALYPTS

There are over 700 species of eucalypts in Australia, the Philipines, Indonesia, and Papua New Guinea, almost all in the genus *Eucalyptus*, but some in closely related genera. Bark variation is remarkable and used to place eucalypt species in distinctive subgroups. For example, the outer bark of rainbow gums is shed in strips, revealing a spectrum of color. Snow gums and ghost gums have white to cream bark; spotted gums have mottled bark shed in thin flakes; and stringybarks have bark with long fibers. Scribbly gums are named for the characteristic bark patterns created by moth larvae. In half-barks, there's an abrupt transition from dark bark on the lower trunk and light-colored bark above.

EUCALYPT FIRE ADAPTATION

Australia is dominated by dry, fire-prone woodlands and savanna. As a result, eucalypts show examples of fire adaptations. Interestingly, molecular genetics shows that eucalypt fire adaptations originated over 60 Mya—much earlier than the arrival of modern humans in Australia (around 50,000 years ago).

↓ In some eucalypts, epicormic sprouts appear along trunks and branches, speeding up recovery after fire.

↓ The stamens (lower image) are often showy in eucalypts. The pistils develop into the distinctive seed-bearing capsules (upper image).

→ In the rainbow eucalyptus (*Eucalyptus deglupta*), the outer bark peels to reveal green bark that ages to red, orange, blue, and purple. Leaves are opposite (in most other eucalypts the adult leaves are alternate). This tree occurs in wetter areas than most eucalypts, and is the only eucalypt found growing natively in the Northern Hemisphere.

BARK AS AN ADAPTATION TO FIRE

Trees adapted to fire are called pyrophytes and have different coping strategies, which reflects the fact that fires themselves vary. One way to quantify the variation is through the concept of fire intensity.

Fire intensity can be represented by the maximum temperatures reached and the durations and vertical distributions of heat exposure. At one extreme are quickly moving "ground" fires that reach temperatures of several hundred degrees, move through the forest understory in minutes, and have little effect below ground or in the canopy. At the other extreme are crown fires that also ignite surface fuels and soil organic matter and which have temperatures that exceed 1,832°F (1,000°C), with heat exposure that lasts hours. Fires can be intermediate between these extremes. Some ground fires produce long-duration heat if they smolder slowly. Some canopy fires move through tree crowns without consuming soil organic matter below.

FIRE-RESISTING STRATEGIES

For low to moderate fire intensity, the best survival strategy for existing trees may be increased bark thickness. At intensities that damage standing trees but don't affect below-ground roots and stems, sprouting is also a key strategy and often faster than reproduction by seed, because root systems and underground stems are intact and have stored energy sources. At fire intensities that kill standing trees but don't sterilize the soil, soil seed banking becomes important. At the highest fire intensities that kill standing trees and dormant seeds in the soil, the strategy shifts to new seed dispersal.

If fire kills stems above ground, some species regenerate by sprouting from the stem base or "root collar," as well as underground roots and stems. These sprouts come from epicormic buds produced by the lateral meristem and emerge through the outer bark.

BARK THICKNESS AS DEFENSE

The thicker the outer bark, the slower hot temperatures penetrate the inner bark and the lower the maximum temperature. For example, the bark thickness of canopy trees with diameters of 20 in/50 cm) varies from $1/4$ in/0.5 cm (fire intolerant) to 2–6 in/5–15 cm or more (fire tolerant). Furthermore, fire-tolerant trees allocate more growth to bark at the base of the trunk to a height of $6 1/2$–13 ft (2–4 m). In humid ecosystems, such as the Amazon basin, species have thin bark and are vulnerable to the introduction of fire by human activities. Long-lived, fire-tolerant trees accumulate very thick bark, including the giant sequoia (30 in/76 cm), coast redwood (24 in/61 cm), and Douglas fir (13 in/33 cm).

BARK TRAITS AFFECTING FIRE INTENSITY

Bark traits affect not only tree survival and regeneration, but also fire intensity itself. For instance, Australian stringybarks have festoons of bark-ribbons that act as "fire ladders," moving the fire from ground level to the canopy. The wafting of burning bark also promotes fire spread. The stringybarks propagate from seeds stored high in the canopy, which germinate after the fire on mineral soils left behind by intense blazes.

↑ The stringybark (*Eucalyptus globoidea*) produces narrow ribbons of bark that allow ground fires to spread upward and reach the tree canopies.

Trees also differ in flammability. Under identical conditions, debris from some trees is initially drier and dries more quickly than others, thereby promoting fire. Trees like eucalypts and many gymnosperms also contain resins and other compounds that promote fire ignition. Other species, such as aspens in North America, slow the movement of fire and reduce fire intensity due to high water content in their leaves and stems.

USES OF BARK

Cork oak (*Quercus suber*) is a drought-tolerant, long-lived, broad-leaved evergreen tree native to southwestern Europe and northwestern Africa. It has also been planted throughout the Mediterranean region. Typical of trees in fire-prone areas, cork oak produces thick bark (potentially 8 in/20 cm, or more), consisting of dead, air-filled cells with cellulose walls that insulate the living inner bark from high temperatures during fires.

Cork oak bark also contains suberin, which is water-repellent, increasing its value as a bottle-stopper. The bark is harvested by hand at about 1¼ in (3 cm) thick. When harvested by hand without injuring the cork cambium, individual trees can be reharvested on a seven- to twelve-year cycle. The best-quality and most uniform cork comes

LEATHER TANNING

Bark tannins are acidic chemicals used in the tanning of leather. Tannins bind to collagen proteins, rendering animal skins water-repellent and less susceptible to breakdown. The bark of hemlock, oaks, chestnut, mangrove, acacia, and other species was harvested and processed in bark mills. In the 1800s, one of the principle values of forest harvest was for tanning.

after the third harvest. The bark is used for bottle corks, sound and heat insulation, flooring, cork boards, the handles of fishing rods and hand tools, and sporting equipment (badminton birdies and cricket balls). Fifty percent of commercial cork comes from Portugal.

OTHER USES OF BARK

Bark has been used in many other ways. Chemicals that evolved as a defense against fungi or herbivores have been used as spices and flavorings (for example, cinnamon and wintergreen) and as medicines (aspirin from willow, quinine from the cinchona, and taxol from yew). "Root" beers were flavored with bark extracts. Resins and latex, produced to poison or repel herbivores or to quickly repair bark damage, are tapped for making products in the naval stores industry and in the manufacture of rubber.

A variety of dyes are also obtained from bark. Bark fibers were used to make cordage in many early societies. The dried inner bark of trees was used to make bread in many parts of the world; after the introduction of flour derived from crops, bark bread became known as a famine bread because it was only used when grain crops had failed.

← The remarkable bark of the cork oak (*Quercus suber*) evolved to insulate the living inner bark of the tree from lethal temperatures produced by fires in fire-prone areas.

UNIQUE ROOT
MACRO-STRUCTURES

There is remarkable variation in root geometry and morphology. However, because tree roots are at least partially below ground, they are very difficult to sample. Excavating roots to make even cursory inspections of how they function is a logistical challenge, since it can damage them—even more so for large trees or those growing in rocky soils.

Tree roots help stabilize trees against windthrow and keep them balanced when wet soils lose their support capacity. We have already seen how tree roots dropping from branches and becoming trunks support the spread of banyans (see page 56).

PANDANUS PROP ROOTS

Pandanus is a genus of about 750 palm-like small trees and shrubs. These trees are of cultural, medicinal, and economic importance in the Pacific, second only to the coconut on atolls. *Pandanus* trees produce prop roots, which develop from the trunk as unbranched aerial roots and grow to form a pyramidal support structure that mechanically supports the trunk.

↓ The pneumatophores of black mangrove (*Avicennia germinans*) in Bermuda protrude from the water surface to aerate the roots below.

↓ Growing in the sandy soils of the Kalahari Desert, the shepherd's tree (*Boscia albitrunca*) has roots that are over 230 ft (70 m) deep.

→ Hala or screw pine (*Pandanus tectorius*) prop roots at the Ho'omaluhia Botanical Garden, O'ahu, on Maui island, Hawai'i.

THE METABOLIC COSTS
OF TREE ROOTS

Usually, the mass of tree roots is about 20 to 25 percent of a tree's total biomass. However, roots can account for 40 to 50 percent of the tree's energy expenditures. The metabolic demands of roots are double that of a comparable biomass of other types of plant tissues. This arises due to the costs of growing and replacing fine roots; the costs of supporting the energy demands of mutualistic interactions with mycorrhiza (see page 86); and the energy needed for the biochemical transfer of nutrients from the soil to inside the tree across root membranes.

TREE ROOT SYSTEMS

As is the case for tree canopies, some regular geometries occur in the larger roots of trees of different species. These patterns in tree roots can be classified by the dominant type of roots forming the root system and the shape of the resulting root mass:

- Taproot systems: Dominated by roots that grow downward from the rootstock. The systems are found in younger or smaller trees. Generally, they are unable to grow more deeply, often because of the lower oxygen levels and lower nutrient levels in deep soils.

- Lateral root systems: Dominated by roots that grow outward from the tree, just below the ground surface. Sinker roots grow downward from the laterals to capture nutrients and water from the deeper soil.

- Heart root systems: Formed of roots that grow obliquely from the tree rootstock to form a large, heart-shaped root ball.

ENERGY FOR MAINTAINING ROOTS

The expenditure of energy to grow new roots and replace dead roots combined with the energy needed for living roots makes the roots the costliest tree tissues to maintain. The maintenance energy for roots is met by the oxidation of sugars produced by photosynthesis. The photosynthetic process uses energy from sunlight to combine water and carbon dioxide to make sugars and produce oxygen. The respiration processes in roots and other tissues oxidize (burn) the sugar to obtain energy, while releasing carbon dioxide in the process.

EFFECT ON ACIDITY LEVELS

When the carbon dioxide released by root respiration processes combines with soil water, it forms carbolic acid. Nutrient uptake by fine roots and the decay of organic matter in dead plant parts also produce acidity. Professor Phillip Sollins and colleagues (of Oregon State University)—in the first study to estimate total acidity production in a forest—found that a Douglas fir forest in western Oregon produced considerable acidity that was almost all consumed by weathering rock. Over 75 percent of this acidity came from root-related processes: root respiration, the chemical uptake of nutrients, and the decomposition of dead roots. The Late Devonian extinction event (see page 16) has revealed the importance of root processes in changing local and global chemistry and in working as possible agents of global change. In that case, the root-derived increase in rock-weathering processes was associated with the dominance of the first rooted trees, *Archaeopteris*, an effect still evident in the roots of modern forests.

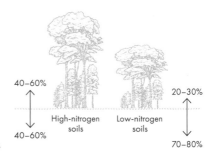

→ The percentage of carbon fixed over a year allocated to wood growth above ground and fine root growth below ground. When soil nitrogen levels are high, 40–60 percent of the production goes to above-ground wood; under low nitrogen this drops to 20–30 percent.

40–60%

40–60%

High-nitrogen soils

Low-nitrogen soils

20–30%

70–80%

MYCORRHIZAE

T he village of Rhynie is located about 40 miles (64 km) north of Aberdeen, in Scotland. It has a population of 138 people and beautiful surroundings. If one turns back time 410 million years, this peaceful location was in the midst of a geothermal hellscape with hot springs of silica-rich water and thermal vents that would flood small ponds and streams. A modern analogue would be the geyser fields found in Yellowstone National Park, in Montana, in the United States.

SYMBIOTIC FUNGI IN THE RHYNIE CHERT

At the Rhynie location, the hot springs rapidly petrified plants, arthropods, fungi, lichens, and algae in a chert (flint) rock called the Rhynie Chert. Sliced microscopically thin, the fossils were preserved in remarkable detail. One could see symbiotic fungi inside the cells called mycorrhizae, which were identical to those found in modern plants. The first land plants were evidenced by fossilized microscopic spores from some 475 Mya, but the incredible detail of the fossils of the Rhynie Chert make this a spectacular find.

Mycorrhiza derives from *myco*, meaning "of fungi," and *rhiza*, which is Greek for "root." The symbiotic fungi (*Glomites rhyniensis*) seen in the Rhynie Chert are endomycorrhiza, in which the fungus penetrates the plant cells and forms feathered structures called arbuscules. These increase exchanges between the fungus and the plant. The endomycorrhizal fungus also forms vesicles, small sacs that store fats to supply the fungus with energy. Strands of fungal hyphae spread through the surrounding soil to obtain water and nutrients for the plant. Endomycorrhiza are also called VA mycorrhiza, in reference to their unique Vesicle and Arbuscule structures. Other symbiotic fungi are ectomycorrhiza, and these have fungal hyphae that colonize roots by pushing the root cells apart but not penetrating them.

ENDOMYCORRHIZAL
OR ECTOMYCORRHIZAL

Trees with finer, small roots that form fibrous mats are more likely to be symbiotic with endomycorrhiza. Thickened and fleshy end-roots are indicative of trees that are more likely to have ectomycorrhizal symbiotic associations with the roots—around 10 percent of plant families, notably orchids and many woody plants such as the birch, dipterocarp, eucalyptus, oak, pine, and rose families.

↓ New leaves have emerged on these beech (*Fagus sylvatica*) trees in the Gribskov Forest, in northern Zealand, Denmark. These trees have several symbiotic mycorrhizal fungi.

THE RHIZOSPHERE

The volume of soil immediately surrounding and under the physical and chemical influence of the roots is called the rhizosphere (*rhiza* is Greek for "root" and *sphere* connotes a zone of influence).

ROOT INTERACTION WITH
THE RHIZOSPHERE

The surfaces of fine roots are sites of chemical and physical processes, which can dramatically change the nature of the adjacent soil. Roots leak sugars, emit carbon dioxide and acids, and take up water. The soils that are microscopically adjacent to active root surfaces tend to be depleted of life-required chemicals. Roots, particularly fine roots, deplete the surrounding soil of nutrients over time. In response to their micro-degradation of the soil, the fine roots of trees have high rates of death and decomposition and are rapidly replaced by new roots to explore "fresh" or at least not recently root-exploited soil.

Fine roots have a root cap that is pushed through the soil as they grow. The root cap is composed of a material that resembles hardened tree bark. After root extension, the new root surface is open for exchanges with the surrounding soil. These roots can undergo suberization, a process that blocks transport in and out of the root. Root hairs can shut down, suberize, and then reinitiate growth, with new roots pushing the root cap through the soil.

LIFE IN THE RHIZOSPHERE

The rhizosphere is thin (around 1 mm thick), remarkably diverse, and a tremendous reservoir of very different microorganisms. Using the number of genes in a given area as an indicator of diversity, most of the genes in the rhizosphere are "owned" by plants and found in the roots. This is followed by the nematode worm gene count, then by the arthropods, protozoa, algae, and bacteria. Embedded in this diversity are symbionts that aid plants in different ways, such as by producing antibiotics. Others are plant pathogenic microorganisms that can colonize the rhizosphere, overcome plant

A FOREST IN MINIATURE

The complexity of the rhizosphere rivals that of the rest of the forest ecosystem. Its microscopic scale makes it difficult to study, but developments in modern biology have set the stage by providing tools for an era of discovery. Today, we need to gain a much greater understanding of the possible effects of transporting soil and plant material between continents, particularly with the contemporaneous changes in climate.

defence mechanisms, and cause disease. Still other microorganisms function as opportunistic human pathogens and cause disease when introduced into those with weakened immune systems.

To the plant, the rhizosphere is a great tangle of friends, allies, enemies, and dangerous liaisons. For example, a particularly invasive species, *Phytophthora cinnamomi*, is a soil-borne, fungus-like organism that produces an infection that causes a condition in plants called "root rot" or "dieback." This was potentially spread by the movement of plants and soils in the time of sailing ships. It, or something much like it, was reported in the United States two centuries ago. The islands of Southeast Asia are considered its likely evolutionary home. It was described in Sumatra in 1922 as a disease of cinnamon (*Cinnamomum verum*). Now found worldwide in 70 countries, this soil pathogen is extremely invasive and infects almost 5,000 plant species. Its tropical origin portends a future increase in its potential range due to global climate warming.

→ Fungal mycelium or hyphae. When mycorrhizae attach to tree roots, their hyphae greatly increase the volume of soil interacting with the tree roots (the rhizosphere).

NITROGEN FIXATION

Some bivalent nitrogen (N_2) can be made available to plants (fixed) by lightning and ultraviolet light, which join the nitrogen with oxygen to form nitric oxide. However, over 90 percent of nitrogen fixation is attributable to the actions of microorganisms. Some of these are free-living in the soil, but others are symbiotic nitrogen-fixing bacteria associated with the roots of cereal grasses and leguminous plants, the latter of which include some tree species. While the relationship between the nitrogen-fixing bacteria associated with the fine roots of plants resembles mycorrhizal symbiosis, it involves bacteria and not fungi. Nitrogen-fixing trees form a symbiosis with nitrogen-fixing bacteria in nodules or swellings in the root hairs.

ELUSIVE COMPOUNDS

Nitrogen compounds are essential chemicals in the biochemical functioning of plants. Crops are often fertilized with nitrogen in the form of ammonium compounds and nitrate compounds. Although the atmosphere is 78 percent nitrogen—plants are bathed in this essential element—it is frequently characterized as the major limiting nutrient in forests. This is because the nitrogen in the atmosphere is almost all bivalent nitrogen (N_2), with two nitrogen atoms bound together by a triple chemical bond. N_2 is a colorless, odorless, tasteless, and inert gas, and it cannot be used by plants.

NITROGEN-FIXING TREES

Nitrogen-fixing trees are self-fertilizing, at least with respect to nitrogen, and have an advantage in low-fertility sites. They are often included in tree plantings to revegetate abused land and improve soil fertility. The trees can grow rapidly on a wide variety of sites, including abandoned agricultural fields. For reclaiming coal strip-mine sites in Appalachia, and for erosion control in Europe and Asia, black locust (*Robinia pseudoacacia*) is a preferred species because of it nitrogen-fixing capacity. The higher nitrogen content of its leaves and stems makes it superior fodder for browsing herbivores (deer and elk)—another benefit of establishing wildlife populations on abused lands. Other nitrogen-fixing trees, notably *Acacia*, similarly support wildlife herds in African game parks. Giraffes are particularly adapted to feed on thorny acacias, using the horny plates in their lips to crush thorns.

← Black locust
(*Robinia pseudoacacia*)
is a superior tree for
reclaiming wasteland
and also provides
flowers for bees.

ROOT-FEEDING INSECTS

U nlike most root-feeders, cicadas are loud, large, and conspicuous. They are members of the order Hemiptera (true bugs) in the superfamily Cicadoidea with two families: the primitive "Hairy Cicadas" (Tettigarctidae) are represented by two species found only in Australia. The other family (Cicadidae) has 3,200 species and a worldwide distribution.

Cicadas have sucking mouthparts, which allow them to penetrate their tubular beaks into trees to feed on tree sap, either on roots as larvae or on above-ground tree parts as adults. They are large insects. The largest is the Malaysian empress cicada (*Megapomponia imperatoria*), which has an 8 in (20 cm) wingspan. Cicadas are eaten by many predators, including people, and are considered a delicacy in some parts of the world. One genus of cicadas (the *Magicicada*) in North America produces broods every 13 or 17 years, depending on the species. The massive eruptions of the species after such long intervals may eliminate specialized but short-lived predators.

↓ Root knot on the Hawaiian *Acacia koa* caused by *Meloidogyne* nematodes. Nematode roundworms induce gall formation (knots) on the roots, weakening the tree.

↓ Root aphids (such as *Prociphilus americanus*) produce several generations of live young each year. Part of the population remains on the host all year round.

→ A pair of 17-year periodical cicadas (*Magicicada*). The population eruptions may thwart some of the effects of predation in a similar way to the "masting" seen in seeds (see page 97).

LIFE HISTORY

The phases of a tree's life—from seed to seedling, to sapling, to mature (reproductive) tree, and back to seed—are, collectively, its life history. Despite sharing the same overall form, tree species, even those growing in the same forest, vary in the timing and nature of their life history events. This variation is one of the keys to understanding the differences between tree species and the pressures that have shaped tree evolution.

Life history characteristics can be correlated with one another, producing syndromes or "life history strategies" that vary among species. This chapter is organized around two of these correlations: first, the smaller the seed, the greater the number of seeds that are produced in an annual seed crop; and, second, the faster the growth rate, the younger the age at first reproduction and the shorter the life span. Although there are exceptions, these generalizations are a starting point for recognizing the ecological differences among trees.

↓ Oak trees produce acorns, nuts with a cap at one end and a hard shell protecting the embryo and cotyledons.

↓ As the oak seedling emerges, it establishes roots, stems, and leaves, thanks to the rich energy sources in the cotyledons.

→ Most of the 450 species of oaks (*Quercus*) are found in the north temperate zone, but some occur at high elevations in the tropics. Major subgroups include the red oak (*Q. rubra*), with deciduous leaves with pointed, bristle-tipped lobes (illustrated here), the white oak (*Q. alba*), with deciduous leaves that have rounded lobes, and the American live oak (*Q. virginiana*), which has unlobed evergreen leaves.

SEEDS AND SEED CROPS

Botanically, a seed has three components: an embryo that forms a fertilized ovule, a tissue called endosperm that often contains an energy supply for germination, and an outer covering called the seed coat. Gymnosperms (mostly needle-leaved trees with cones) are seed plants but lack true flowers. In angiosperms (flowering plants), the ovule is enclosed within an ovary. The ovary and other floral tissues develop into the fruits that contain the seeds.

SEED SIZES

Tree seeds span a huge range of sizes, from the cottonwood (*Populus species*) at 0.000004 oz (0.0011 g) to the coco de mer (*Lodoicea maldivica*), a relative of the coconut palm (*Cocos nucifera*), at 44 lb (20 kg), an astounding 15-millionfold difference. The large seeds of coco de mer and coconut palms are necessary because both trees live on tropical shorelines and their seeds must be provisioned by the parent tree with a rich energy source to ensure germination and early survival on sands that have sunlight but lack a developed

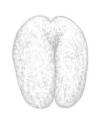

↑ The world's heaviest seed, the coco de mer (*Lodoicea maldivica*), is contained in a thick husk that reaches a diameter of 20 in (50 cm).

SEED NUMBERS

Seed weight is usually correlated with seed crop size. A mature coco de mer releases 50 seeds a year, but a cottonwood can release over 25 million seeds a year. The small seeds have a cottony web and are transported by wind over many miles. While coco de mer seeds reflect the challenge of early seedling survival, the cottonwood seeds reflect a strategy of finding recent disturbance patches with high enough resource levels for germination.

soil. The coconut combines this provisioning of the seed with the ability to withstand dispersal on salty seas. In contrast, a cottonwood seed has little parental investment in early survival, suggesting these seeds need bright light and soils with high resource levels, such as the alluvial soils produced by erosion and deposition along riversides.

THE ACORN

In the temperate zone, the oaks (*Quercus*) comprise about 450 species characterized by the acorn. Although less extreme than the coco de mer, the acorn also represents parental investment. We can surmise from the acorn that the seedling will also face early challenges, in this case, shade and thick leaf litter that the roots must penetrate before the seedling becomes self-supporting through photosynthesis.

The acorn's energy supply attracts animal foragers (wildlife, insects, and, at least historically, humans and livestock), so defensive strategies are needed. The oak uses two. First, the acorns contain tannins, chemicals that interfere with protein function, though some animal species are tolerant (acorns were leached of tannins for human consumption). Second, oaks produce heavy crops every three to five or more years, with scarcity in intervening years, a trait called "masting."

MASTING

Masting results in a year of plenty followed by several years of famine. This, in turn, prevents animal populations from increasing to the point of complete consumption of the crop of acorns and other fruits in years of large production, a phenomenon known as "predator satiation." In addition, birds and mammals become dispersal agents for acorns that are not consumed.

Masting is synchronized over tens to hundreds of miles, although it isn't fully understood how this occurs. In North America, the now-extinct passenger pigeon evolved a strategy to take advantage of the masting defense: their flocks were among the largest that have ever occurred and roamed over large distances in search of the shifting location of large acorn crops.

SEED DISPERSAL

Seeds and fruits are adapted to various dispersal vectors. Some wind-dispersed seeds follow the cottonwood model (see page 96), while others have "winged" seeds that whirl in a helicopter-like fashion from treetops, slowing seed descent, so seeds are dispersed farther from the parent tree.

Tree seeds can also be dispersed by gravity, water, and animals. Some seeds are even scattered by "explosive dehiscence." An extreme example is the sandbox tree (*Hura crepitans*) of the American tropics: on dry, sunny days, the capsules open with a loud "pistol crack," expelling seeds at over 124 mph (200 km/h) and reaching distances of over 131 ft (40 m). The flowers and thus the fruits are borne directly on the tree's trunk, an unusual trait called "cauliflory" that is shared with cacao trees, the source of chocolate.

ROTTING FRUIT ENIGMA

In 1982, ecologists Dan Janzen and Paul Martin studied the enigma of rotting fruit in Costa Rica, which lay beneath trees without being dispersed by animals. They identified about 30 woody plant species that seemed to lack dispersal agents. They conjectured that this was the result of the extinction of large mammalian herbivores, like giant ground sloths, around the time humans first appeared after the Pleistocene Period. The same logic has been applied to such species as the avocado and to a number of temperate trees like the Osage orange, with its large, tough-skinned fruits, which suggest consumption and dispersal by extinct mastodons and extinct species of horses and camels.

HITCHING A RIDE

Passive animal transport occurs for seeds and fruits with hooks and sticky coverings, but many animal species transport seeds actively. Ant-dispersed seeds carry a special fat body (aril) attractive to ants, a structure that, per weight of seed, requires a lower energetic investment than the bright, often red, fleshy fruits attractive to birds and which match the gape size of bird beaks. Bat-dispersed fruits are drab (bats are colorblind), have musty or rancid odors, and are juicy (bats have blunt molars to press out the juice, while expelling seeds). For bird- and mammal-dispersed fruits, germination can be enhanced because digestive acids erode the hard seed coat, allowing water and oxygen to penetrate after the seed is defecated.

← The 16 carpels of the sandbox tree are arranged in a ring. As the fruits dry, they pass a critical threshold, leading to an explosive release of the seeds.

SEED DORMANCY

T he seeds of some species are ready for germination when they are shed, while others are in an arrested state of development called dormancy and will not germinate in response to good environmental conditions until their innate dormancy is "broken." For example, in seasonally cold environments, seeds that develop in the fall may lie dormant until the spring; exposure to winter cold or lengthening daylight may be required to break their dormancy. Generally, tropical forests, with year-round warmth and moisture, are characterized by non-dormant seeds. In seasonal, drier, and colder environments, dormancy is common.

DORMANCY TIMES

Some species have much longer dormancy, from a few years to many decades. The seeds of the pin cherry (*Prunus pensylvanica*) in North American forests can remain dormant in the soil for decades. After a forest disturbance, ecologist Peter Marks showed that soil nitrate levels increase, triggering germination. The dormant seeds

FACILITATING GERMINATION

In dormant seeds, the seed coat can be a formidable barrier to water and oxygen and must be breached for germination to occur. Some seeds benefit by passing through the digestive tracts of birds or mammals. Gardeners can prompt germination by mimicking natural factors, such as exposing seeds to cold, heat, light, changing day length, hot water, or acidic solutions, or by scratching the surface with a steel file.

---

of pin cherry and other species accumulate in the soil as a "seed bank," a strategy for germination after forest disturbance, though they can be lost to soil erosion and extreme fires hot enough to sterilize the soil.

The retention of viability in dormant seeds can be astounding. A date palm (*Phoenix dactylifera*) seed recovered from King Herod's tomb germinated at 2,000 years old. Seeds more than 30,000 years old have been recovered from glacial deposits. While these seeds were damaged, they contained tissues that were used to regenerate living plants.

VEGETATIVE REGENERATION

Some trees also regenerate vegetatively. This can happen through cloning or sprouting from the stem base, trunk, or branches after injury to the main stem. Sprouting has advantages over seed production in that the root system is well-developed and may store carbohydrates. Hence, sprouts are often able to grow faster than seedlings. A major disadvantage is that sprouting by-passes sexual recombination and the genetic novelties on which evolution ultimately depends.

← Date palms are cultivated in Africa, the Middle East, and South Asia. Individual trees are male or female. Though naturally wind-pollinated, plantation trees are hand-pollinated so that more female date palms can be planted.

DEVIL'S WALKING STICK

The North American devil's walking stick (*Aralia spinosa*) has a remarkable growth form. In the "trunk-building phase," branchless stems quickly grow upward, the leaf display consisting of sets of 3-ft (1-m) long, twice-compound leaves. After two to five years and at a height of about 10 ft (3 m), the stems produce a terminal inflorescence. The tree then enters a "crown-building phase," with branches developing from the lateral buds below the inflorescence and spreading leaves away from the main axis.

In recent treefall gaps, the ecological advantage goes to species that can colonize and grow quickly. *Aralia* height growth exceeds 3 ft (1 m) per year, two to five times greater than shade-adapted species in the same forest. This comes at a cost: due to the lower investment in wood strength, *Aralia* stems survive only about 25 years, less than one-tenth the life span of the shade-adapted species (like beech). An early onset of reproduction compensates for this short life span. Devil's walking stick first bears seed at two to five years, compared to up to 40 years in beech.

↓ Young devil's walking sticks have prickles to deter mammals. These are gradually shed and replaced with thickening bark as the tree matures.

↓ A single, doubly compound devil's walking stick leaf may reach a yard (1 m) in length and consists of 50 or more leaflets.

→ In its early years, the devil's walking stick maximizes height growth by using large, compound leaves, rather than constructing a series of stem branches needed for smaller leaves. When the plant is about 12 ft (4 m) tall, the terminal stem bud produces an inflorescence, after which lateral buds create a branched crown.

GROWTH RATE AND LIFE STAGES

Devil's walking stick (named for the thorny, unbranched stems of the trunk-building phase) is at one extreme of tree life histories—fast growth, early reproduction, and short life span. By "growth rate," we mean the rate of the tree frame's three-dimensional spread rather than the rate of carbon gain. This upward and outward extension of the terminal meristems is called primary growth, in contrast to the increase in diameter known as secondary growth, the latter produced by the lateral meristem or cambium (see page 18). The growth rate varies within a species as a function of environmental conditions, but it also varies greatly across species.

GROWTH AND WOOD DENSITY

One way to grow faster in terms of three-dimensional development is to produce wood with a lower weight per volume—thus a lower wood density. Wood density has proven a useful index for life history strategies. It varies with tree age and environment, but here we are interested in the differences among species. In the tropics, species of *Cecropia* and balsa wood represent the extreme of fast-growth, low-density species like the devil's walking stick in temperate forest. At the other extreme, ebony trees in the tropics and beech trees in the temperate zone represent the extreme of slow growth and high-density wood—so high in some species that the wood sinks in water.

↓ Like devil's walking sticks, *Cecropia* maximizes height through large leaves, but these are palmately lobed rather than compound.

LIFE HISTORY SELECTION

Ecologists have theorized about the selective forces behind life history strategies. Species that are r-selected are those at the fast growth/early reproduction/short life span end of the spectrum (the "r" is the population growth rate parameter in population models). These species have fast population growth rates after forest disturbances when resources are high and competition low but are less competitive as forests mature. K-selected species are at the slow growth/late reproduction/long life span end of the spectrum ("K" is the parameter for carrying capacity in population models). They are strong competitors that increase in dominance with time-since-disturbance. The replacement of r-species by K-species is one way of thinking about succession after disturbance (see page 112). The r-K perspective also embeds the idea of trade-offs—that no species can do everything well; hence, individual species become specialized for different environmental conditions and successional roles.

GRIME'S TRIANGULAR MODEL

Phillip Grime created a model for plant strategies with three axes: disturbance, competition, and stress. Here, the extreme r-selected species were adapted to disturbances that occurred in rapid sequence, thus selecting for early reproduction. He called these "ruderal" species. The extreme K species were termed stress-tolerators because they dominated sites that were always stressful (perhaps due to droughtiness or low nutrients) or became stressful as available resources decreased in late successional forests. Grime's competitive species were those that dominated areas of higher resource availability and lower rates of disturbance than the ruderal species, producing longer life spans and delayed reproduction. Whereas ruderals dominate very early in succession, competitive species (in Grime's sense) dominate from early succession to late successional stands but are ultimately replaced by stress-tolerators. In addition to their increase in late succession, stress-tolerators are also successful on environmentally stressful sites.

The r-K selection scheme and Grime's triangular model offer explanations of life history strategies, but they are also, along with environment and species interactions, ways of defining the unique niches occupied by different tree species.

PRIMARY AND SECONDARY SUCCESSION

Ecological succession involves the patterns of change in vegetation over time at a particular location. In the case of primary succession, a previously unvegetated new surface, such as recently exuded and cooled lava, a newly formed sandbar in a river, or the exposed rock at the foot of a receding glacier, is colonized by different kinds of plants. As soil develops on the site, trees may come to dominate and forests develop. In secondary succession, a disturbance (wildfire, flood, hurricane, tornado) eliminates or severely damages the vegetation, but the soil under a previously existing forest remains largely intact. In successions involving trees and forests, secondary succession generally proceeds more rapidly than primary succession to forest conditions.

In forests, the changes associated with succession occur more rapidly early in the process. In older forested sites, change may occur so slowly that it is difficult to detect on human time scales and the ecosystem would appear to be in equilibrium.

↓ Primary succession of lichen to mature trees: (A) lichens produce weak acids that eat away at the granite, disintegrating the surface to form a shallow soil, (B) chemicals from the soil, rainwater, and wind accelerates erosion, (C) eroded soil collects in low parts, allowing shrubs to grow, (D) trees take root downslope where granite-derived soils have accumulated in greater depths.

→ The flowering evergreen tree 'ōhi'a lehua (*Metrosideros polymorpha*) is shown here colonizing an over 30-year-old lava flow from the Kīlauea volcano located at Kalapana in the Puna District of Hawai'i.

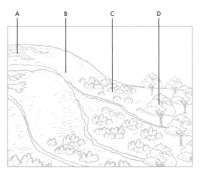

GAP REPLACEMENT

There have been some remarkable reconstructions of successional changes in forests. Landscape processes can help us solve the puzzle by forming chronosequences, as in these examples:

- Sequentially younger vegetation hiking up the valley of a receding glacier, in Glacier Bay, Alaska.
- The development of soils and vegetation from a coastline sequence, in Indiana Dunes National Park.
- In the changes in forests growing on different-aged lava flows on the island of Hawai'i.

REGENERATION IN AN OLD FOREST

Hike through an old forest and one can observe the interactions of ecological processes generating a forest pattern. Begin with a close inspection of a large fallen tree, perhaps toppled in a windstorm. The space left by the toppled rootstock is now a hole in the soil, perhaps wet or even with standing water. The tipped-up root mass will have torn mineral soil from below the leaf and soil layer of the forest. The hummocks and hollows in the soil created by this process are indicators of relatively undisturbed sites. Some tree species require exposed soil to establish new saplings. The fallen tree trunk can also be a regeneration site for other species.

The gap in the forest canopy allows for increased available light. If a canopy gap is large, one might see patterns in regenerating plants, as the western side of the gap gets morning sun when it is cooler and more humid, which contrasts with the evening sun in the hotter afternoon on the eastern side. The fall of the former canopy results in a site of destruction and exposed mineral soil.

The development and diversity of the forest can be seen as a mosaic of the generation and recovery after treefall. The process maintains diversity. The fall of one species of large tree generates regeneration opportunities for many trees of many species.

SUCCESSION PATTERN PUZZLE

Studying succession patterns is like solving a jigsaw puzzle—a landscape patch representing the succession process at a point in time is "fitted" to similar pieces, some older and some younger, to reconstruct the longer succession pattern. Pieces may be missing, since there may not be any patches of forest arising from events in particular age ranges. Or there could be extra pieces from another puzzle: slight differences in soil can produce differences in patches of the same age.

↓ This eastern hemlock (*Tsuga canadensis*) began its life growing on a fallen log. When the log eventually rotted away, the hemlock tree was left perched on a series of stilt roots.

THE FOREST AS A DYNAMIC MOSAIC

T he processes involved in forest change at the scale of gaps in the forest canopy are part of a grand ecological paradigm eloquently presented by A. S. Watt in his 1947 presidential address to the British Ecological Society. If one could read only one paper on vegetation ecology, then his 1947 "Pattern and process in the plant community" would be it.

INTERACTION OF FOREST
PROCESSES AND PATTERNS

Processes in a forest can produce patterns—patterns in the physical structure of a forest, in its most successful species, or in the numbers of smaller trees compared to large trees. A forest dominated by large trees will have larger gaps that allow higher levels of sunlight to reach the ground. In the tropics, such increased high sunlight produces hotter, drier forest-floor conditions that can be lethal to the seedlings and saplings already growing in the shady conditions under the canopy. Other species, adapted for dispersal, occupy the areas under the canopy gaps. One might expect these species to have physiologies that allow them to tolerate gap micro-environments in their gamble for existence.

Patterns in a forest can also influence processes. In the example given for forest processes above, what if the dominant trees in the forest were smaller and produced smaller canopy gaps? Then, the seedling and sapling trees struggling to survive in the shade of the forest floor would not be killed by the change in physical processes on the forest floor from gap formation overhead. Indeed, they might be able to prosper while inching up into the canopy. The Dutch tropical ecologist Roelof A. A. Oldeman proposed that tropical forest trees should be regarded as either "strugglers" or "gamblers" to appreciate the dynamics of the rainforest.

The paper describes a unification of processes (seed germination, growth, mortality from age or from events, and so on) interacting to produce the patterns one sees in heathlands, grasslands, bogs, alpine shrublands, and forests. In his doctoral work published in 1925, A. S. Watt had noted that on his study sites, ash (*Fraxinus excelsior*) seedlings grew in the sunny centers of canopy gaps while beech (*Fagus sylvatica*) regeneration took place in a more shaded "circle of regeneration" at the shaded edges of the gap.

DIFFERING COMPOSITIONS AND STRUCTURES

Often, but certainly not always, in more zoological studies of communities, negative interactions are important drivers of community change. Predators reduce prey numbers or similar species compete against each another. Trees are so large that they strongly influence their environment at the micro-scale and even at large-space scales. Large patches of forest in a non-forest matrix—think of forest patches in an agricultural landscape as an example—are different in composition. So much so that two landscapes with the same area of forests, but arranged with a few large patches in the one case and many small patches in the other, support forests with a different composition and structure. The percentage of woodlots in edges versus shaded interiors drives these differences. However, when the composition and structure are different, pattern and process interactions can further change the structure. For this reason, managing forest landscapes for conservation or diversity protection requires attention to the geometry of forests as well as to the acreage of forest land.

→ The fall of a large tree opens the canopy of a forest. On the ground, there is destruction in the area where the tree canopy fell, a raised site that some plant species prefer for regeneration, an elevated mixture of soil and roots in the tipped-up root ball, and a hole in the forest floor.

SECONDARY SUCCESSION: PONDEROSA FIRE RECOVERY

Secondary succession occurs after a disturbance event (fire, clearcutting, flood, farmland abandonment, or avalanche, for example) destroys the vegetation at a location and recovery processes are initiated. In secondary succession, there is a "heritage" from what came before in the remaining soil or in sources of seeds or plants to regenerate vegetation. Disturbances, either natural or human-made, are part of modern forest landscapes.

PONDEROSA PINE FORESTS

Ponderosa pine (*Pinus ponderosa*) forms extensive forests in western North America. The species regenerates exclusively from seed. Wildfire removes competing ground vegetation and exposes mineral soil, which provides the ideal seedbed. The pine is most flammable in spring when the old needles are dry. Eruptions of stand-killing bark beetles can create landscapes full of highly flammable dead and dry trees with canopies of dead dry needles. Thus, regeneration is fire-dependent. Fire, followed by exposed soils, then by grassy meadows to a Ponderosa pine forest, represents a typical secondary succession pattern for the recovery of the species.

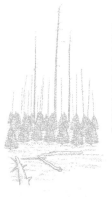

↙ Ponderosa pine (*Pinus ponderosa*) demonstrates obligate regeneration, with wildfires as the trigger. Here, the regenerating seedlings are the same age, while the dead burned trunks are left from the previous regenerative wildfire.

→ Ponderosa pines have their best regeneration in hot wildfires. Born from pyric destruction, they are the most widely distributed pine across North America.

FIR WAVES

We have seen how working out the patterns of forest dynamics resembles solving an ecological jigsaw puzzle. In what is termed "space-for-time" substitution, one attempts to formulate the patterns of forest (or landscape) dynamic change by assembling the pattern of change over time and assuming similar elements of a landscape can be arranged in temporal order.

The patches of a landscape with 60-year-old pine trees surely turn into something like the current patches with 80-year-old pine trees, and the seedling oaks under these pines must presage a mixed oak–pine forest to come. While this may be sensible, there are still problems. Differences in soils, yearly weather differences, fluctuations in grazing herbivore populations eliminating seedlings in one decade, but not another, for example, can disrupt the processes producing the patterns. Other evidence that the underlying dynamics are as one postulates is wonderful to have. Here we discuss a case that does so by displaying pattern regularities: fir waves.

REACHING FOR THE LIGHT

If one imagines the replacement processes from the formation of a canopy gap as a series of images, the first image might show a forest full of regeneration, with seeds germinating and small trees that were saplings before the gap formed (advanced regeneration) racing seedlings and other saplings to be the dominant large tree filling the gap. One can piece this together by noting the degree of recovery or gap closure.

FIR WAVE PHENOMENON

Fir waves can be seen in conifer forests in mountainous regions, notably in New England, in the United States, and Japan. In fir waves, a forest-edge naturally propagates across a forested landscape—driven by the prevailing wind and the associated hoarfrost wind brings. The edge-trees topple and the trees behind them, which were sheltered by these now-dead trees, become the next targets of wind/frost damage. A wave of death travels across the forest. The dead trees, both standing and downed, present as white, debarked trees and logs, which form white stripes. Sheltering among the logs, seedlings germinate and grow. This produces a following wave of birth to form forest. If forest gap dynamics is a large-tree death event followed by a growing-tree recovery, then the fir wave lays this sequence in its natural order before our eyes.

← Balsam fir (*Abies balsamea*) waves in Maine, in the USA. The white bands are dead and downed trees that form distinctive wave patterns across the landscape.

FOREST AREA AND DYNAMIC MOSAICS

There is evidence from satellite-based remote sensing in Canada and Siberia that the area of forests burned averages 2 to 2.5 percent per year. The total percentage and average size is increasing in both places each year. Other forests, notably drier ones like woodlands and savannahs, show similar increases. These changes seem related to global warming.

EFFECTS ON PLANT AND ANIMAL DIVERSITY

The size of the patches making up a dynamically changing forest mosaic and the manner in which the patches are generated over time can have profound implications for the diversity of species in an area of forest. A forest landscape driven by disturbances becomes a patchwork of patches of different sizes. If the dominant disturbances are large treefalls followed by gap-filling processes, then the granularity of the landscape is at gap-scale—perhaps 33–98 ft (10–30 m). If the driving disturbances are small wildfires, the scale of granularity shifts to scales of hectares or square miles, and even larger for large wildfires. Such changes in scale can affect a forest's role as a repository for a diversity of plant and animal species.

As disturbances increase in frequency and size across an area, the proportion of large, old trees diminishes and the landscape elements of the area become younger on average. Species that require a regenerating supply of "old" forest elements to remain in a landscape—for example, those that live in very large holes in senescing trees—are difficult to maintain under higher frequencies of larger disturbances.

The species on the list of the endangered species of the world's forests often have behaviors that demand microhabitats generated over long periods of time in a forest landscape. Trees, old and large enough to be used as nests, take time to develop and forest clearing is reducing the size of landscape available.

The ivory-billed woodpecker (*Campephilus principalis*), a likely extinct large woodpecker of the floodplain forests of the great rivers of the southern United States, is an excellent example of a species that was highly dependent on long-to-develop feeding sites, which would become less frequent under elevated disturbance and/or forest clearing. As painted by John James Audubon in his *Birds of America*, three ivory-billed woodpeckers are shown feeding on wood beetles on a dead, lichen-covered tree limb, removing pieces of bark, but not tearing into the wood. This habit of bark-flaking was regularly reported as the feeding behavior of the species.

Once a woodpecker had removed all the bark from a dead tree, another standing dead tree with loose bark and available insects was needed as a replacement feeding location.

How many such large dead snags does a floodplain forest contain? At what rate are these replaced so that an ivory-billed woodpecker can have a steady supply of places to feed? Unfortunately, such snags appear rarely. A dead tree suitable for use as an ivory-billed woodpecker's feeding station occurs 200 to 400 years after the tree's birth as a seedling, its growth to maturity, and its mode of death. Only when a large, mature tree dies while still standing will the resultant snag be useable by a feeding specialist such as the ivory-billed woodpecker, and then only for a year or two.

→ An adult and two immature ivory-billed woodpeckers flicking loose bark from a dead tree to feed (image derived from J. J. Audubon's *Birds of America*, 1829).

CONSERVATION CHALLENGES

In 2021, the Botanic Garden Conservation International (BGCI) estimated that, of more than 58,000 tree species evaluated, 30 percent (some 17,000 species) were at risk of extinction due to ongoing habitat loss and emerging challenges like climate change and extreme weather events. Using data from the International Union for Conservation of Nature (IUCN) and other sources, BGCI also reported that 440 tree species were on the brink of immediate extinction because fewer than 50 individuals were known to exist and that 142 tree species were likely already extinct. Tree species on isolated tropical islands are particularly threatened because they are often globally restricted ("endemic") to these islands.

NARROW ENDEMICS

While some trees span continents, such as the quaking aspen, which is thought to have the widest range of any North American tree, others are native only to very small areas. Small-range species are called "narrow endemics." The Biota of North American Program uses ranges less than 5,800 square miles (15,500 km^2) to define narrow endemics, but this category ranges downward to examples like *Franklinia*, originally found in a tiny area of 2 or 3 acres in the southeastern United States. Differences in range size likely derive from past evolutionary events and climate variation. Range sizes, along with population sizes, population trends, and direct threats, are among the ranking criteria used by the IUCN's Red List of Threatened Species.

CHALLENGES FACING TODAY'S TREES

Tree species face five challenges in today's world: habitat loss and fragmentation due to conversion of forest to agriculture, other commercial purposes, and urbanization; invasions of species from other continents, including damaging insects and diseases; changes to the ecological processes that influence forests, like fires and floods; climate change and other results of atmospheric pollution; and unsustainable forest harvest. The problem with these challenges is not just the kind of forces at work—after all, some of these phenomena, such as fire, flood, drought, disease, and environmental change, were also present before human impacts—rather, what is new is that the magnitude, rate, and interaction of these forces have increased.

In this chapter, we more fully describe the challenges that conservationists face and the responses that are needed and which, indeed, are already underway.

← Wollemi pine (*Wollemia nobilis*) is a member of the Araucariaceae, a once-widespread family that is now confined to a remote area of New South Wales. It is ranked as critically endangered.

HABITAT LOSS AND FRAGMENTATION

The earliest threat from human populations was the displacement of forests by settlement and agriculture. This displacement caused early extinctions on some islands (for example, those following the arrival of Polynesians on the Hawaiian Islands about 1000 AD), but, of course, the impact became greater as human populations and economies expanded, especially after the Industrial Revolution.

Today, forests are being lost not just to crops and pastureland but also to urbanization, energy extraction, mining and other commercial developments, and monospecific forest plantations. The amount and timing of forest loss has varied geographically. For instance, the boreal forest has been relatively stable in area in recent decades and North American temperate forests have generally increased since the mid-1800s (though not beyond their extent before cutting), but tropical forests are suffering high losses. When habitat is lost, the remaining habitat becomes fragmented, leading to reduced habitat quality, increased edge effects, smaller population sizes, and an interruption to gene flow among populations.

↓ American beech (*Fagus grandifolia*) is a long-lived, slow-growing tree characteristic of old and shady forests.

↓ Gray birch (*Betula populifolia*) is a shade-intolerant tree found at forest edges.

→ Unbroken tropical forest once covered vast areas, with only scattered disturbances from wind and flood and the villages of indigenous peoples. But large-scale forest clearing has produced a dramatic change, reducing habitat and creating extensive areas of forest edge that decrease habitat quality for the species of deep forest interiors.

SPECIES INVASIONS

The individual continents have acted as separate centers of evolution. As a result, 99 percent of tree species are confined to just one continent. In the last several centuries especially, human activities, both intentional and accidental, are allowing species to cross oceanic barriers they could not cross on their own. Species arriving in a new location sometimes do not expand from sites of introduction, but occasionally their populations expand quickly. One explanation for species invasions is called the enemy release hypothesis. This posits that newly arrived species increase quickly because they have escaped the effects of pests and diseases that controlled their populations in their original ranges.

DETRIMENTAL EFFECTS

New invaders often reduce the populations of indigenous species through competition. While this may result in species extinctions in the long run, competition is usually a slow process. Even more concerning are invasions that change natural ecological processes or are accompanied by the invasion of pests and diseases that attack indigenous species.

INVADERS CHANGE NATURAL PROCESSES

Some invading species alter ecosystem processes. For instance, invading trees in the Florida Everglades transpire more water into the atmosphere than indigenous trees, thereby lowering water tables, causing more extreme droughts, and increasing fire severity. A second example is the invasion of Hawaiian ecosystems by nitrogen-fixer plants. Increased nitrogen levels favor fast-growing species that overtop more slowly growing species and smaller individuals.

INVASIVE PESTS AND DISEASES

Invasions can also have rapid impacts through the arrival of new pests and diseases. At the center of this phenomenon is the idea that, when pests and diseases evolve with hosts over long time periods, they will evolve to be less virulent and/or the host will evolve to be more resistant. This "coevolution" potentially stabilizes the host–enemy interaction and the impacts become less severe.

The invasion of pests and diseases into new areas often involves host species that have not evolved resistance. These invasions are particularly severe when the enemies in question are specific to a genus that occurs on both continents. For instance, chestnuts (*Castanea*), ashes (*Fraxinus*), firs (*Abies*), and hemlocks (*Tsuga*) occur in North America and East Asia. In North America, each of these four genera is threatened by pests and diseases accidentally introduced from the Asian continent. Chestnut blight eliminated almost all American chestnut trees between the early 1900s and about 1950 (though root systems continue to produce short-lived basal sprouts). The emerald ash borer is currently causing heavy mortality in ash populations. The balsam woolly adelgid caused almost complete mortality of Fraser fir, an endemic in the southern Appalachian mountains, between about 1960 and 2000. The hemlock woolly adelgid has converted healthy hemlock forests to "ghost forests" from the 1950s to today.

↓ Trees and shrubs in the genus *Miconia* form impenetrable thickets and outcompete indigenous species in many tropical forests across the world.

GHOST FORESTS

Some of the threats to forests act over many decades, but others cause relatively sudden waves of tree mortality. For instance, some trees species are sensitive to increased flooding, salt-water intrusion with sea level rise, severe droughts, extreme fires, and new pests and diseases. Groups of standing dead trees have increased in recent decades and have been called "ghost forests." These have become a strong visual image of the challenges to trees that are underway.

CHANGE IN ECOLOGICAL PROCESSES

Even when forests themselves have not been converted to agriculture or other developments, threats to tree diversity can occur in remnant forests because of changes to the natural processes that play roles in the composition and structure of these forests. Two key examples are the disruption of trophic structures and alterations of natural disturbance regimes.

TROPHIC CASCADES

The loss of large predators, such as wolves, panthers, and large birds of prey, due to eradication programs and habitat loss and fragmentation, can produce cascading effects through the trophic structure of forests. The absence of large predators results in an increase in the herbivore populations that were the prey of those predators. Increased herbivore populations result in the overgrazing of seedlings, understory tree stems, shrubs, and wildflowers. Even a reduction of smaller predators, like birds, can increase herbivorous insects and, thus, increase the defoliation of tree canopies.

FIRE, FLOOD, AND OTHER NATURAL DISTURBANCES

From the beginning, human activities have altered or attempted to alter disturbance regimes, especially the occurrence of fires and floods. Fire suppression has reduced the number and size of fires, but in some forests, this has increased understory fuel levels such that subsequent fires, especially after severe droughts, are more severe. Human activities have also increased fire ignitions through accidental and arson-set fires and through the presence of power lines. Similarly, dams and channelization have reduced flooding as part of the natural dynamics of forests. Conversely, human activity has also increased flooding through the increase in impervious surfaces across watersheds, upstream channelization, and the removal of floodplain forests.

ISLANDS IN A TROPICAL LAKE

Ecologist John Terborgh and an international team recognized an opportunity when a dam was constructed along a river in Venezuela, creating a lake, Lago Guri. What had formerly been hilltops in a forested landscape became islands in this vast lake. The islands were too small to sustain populations of large predators. In a classic trophic cascade, herbivores increased tenfold to hundredfold. Trees declined and ultimately shifted to species with the toughest or most poisonous leaves. Diversity and productivity crashed on the islands. The implications of this "experiment" were clear: nature preserves that are not large enough to sustain trophic structure will suffer continued losses of biodiversity.

↓ An aerial view of a deer-exclusion fence installed to study the effect of deer (excluded to the left, present on the right). The area to the right shows the dramatic effects of browsing.

POLLUTION AND CLIMATE CHANGE

F orests are threatened by two consequences of fossil fuel use: acidification due to the deposition oxides of nitrogen and sulfur, and climate warming from the accumulation of carbon dioxide.

ACID DEPOSITION

Acid deposition is most frequent in the developed and developing countries of the Northern Hemisphere. Nitric and sulfuric acids leach nutrients from tree leaves and soils, potentially reducing productivity in the long term and, in some ecosystems, having other physiological effects. Many developed countries have reduced the use of high sulfur coal in recent decades, and acid deposition has decreased.

CLIMATE WARMING

Precipitation and temperature affect tree distributions, so climate change poses a major concern. Warmer temperatures impact the carbon and water balances of trees (see page 44). Trees are also directly impacted by climate change through an increase in extreme weather events (flood, drought) and fire severity, as well as by sea level rise. Trees are indirectly impacted when interacting species, like pollinators and seed dispersers, are themselves affected, perhaps differently than the trees with which they interact.

↓ Drought has caused a zone of dead tissue around the margin of this sugar maple (*Acer saccharum*) leaf.

↓ Ozone has caused a fine stippling of necrotic tissue throughout this tulip tree (*Liriodendron tulipifera*) leaf.

→ Sea level rise is altering coastlines, reducing island area, and threatening the world's island nations. While these changes may be dramatic where ocean waters and land meet, other effects, such as the salinization of groundwater and the loss of human water supplies, are also the result.

FUTURE FORESTS

A wide variety of resources are harvested from forests, including wood, fuel, fiber, wildlife, fish, water supplies, fungi, fruits, and organisms with medicinal value. It is self-evident that the first goal of forest management should be that these resources are harvested sustainably. Overharvesting of trees can result in soil erosion, the reduction of seed trees for future generations, and poor regeneration.

While strict land protection is critical, the different threats to biodiversity have necessitated varied conservation approaches. Trees play an important role in removing carbon dioxide from the atmosphere and so carbon storage is, itself, a powerful argument for forest protection, particularly of old-growth forests that have accumulated carbon over centuries. However, even protected lands are affected by species invasions, altered natural processes, pollutant deposition, and climate change.

In terms of the extinction of tree species, the global assessment of trees by BGCI found cause for hope, noting that 64 percent of threatened tree species occurred in at least one protected area and 30 percent were present in botanical gardens or seed banks. The report recommended expanding both on-site (protected area) and off-site (gardens and seed banks) conservation.

SAVING SEED

Seed banks and gardens are needed as a last resort against extinction in the wild, to augment wild populations, and restore species to natural habitats. However, tropical trees, which comprise 75 percent of all tree species and are often rare, usually have seeds that are difficult to maintain in seed banks due to a lack of innate dormancy (research is currently attempting to develop better storage methods for these species).

MINIMIZING SPECIES INVASIONS

Proscriptions to avoid or minimize species invasions include restricting the import of species through risk assessment, inspection and monitoring at points of entry, exploration of biocontrol agents for established pests (along with an assessment of the risk of introducing new problem species), and breeding programs and/or gene engineering to increase genetic resistance in trees under threat. Since the movement of some pests has been accelerated by horticultural distribution and by the movement of plant parts, such as firewood, interrupting those movements is critical.

HOW WILL TREES RESPOND TO CLIMATE CHANGE?

In response to the climate changes of the past, tree populations have shifted geographically. A critical question over the coming decades is whether species have high enough dispersal abilities to adjust to environmental changes, a problem compounded by the habitat loss and fragmentation of the landscapes through which they must move. In some cases, at least, the answer is "no" and, therefore, "assisted migration" has been proposed as part of the conservation solution.

MANAGING ECOSYSTEMS

We end this chapter with the concept of ecosystem management, first developed in North America. While protected lands are critical, the sustainable management of harvested lands can contribute to conservation, as can reforestation and the restoration of degraded lands. Ecosystem management is a holistic and regional approach that addresses the restoration of natural processes, the reduction of species invasion, and the need for both large, protected areas and corridors and connections between conserved lands.

→ *Franklinia*, a small deciduous tree in the tea family, has not been seen in the wild since 1803 but has been propagated by gardeners, providing a clear example of "assisted migration."

THE SACRED BODHI TREE

The sacred fig tree (*Ficus religiosa*), also known as the Bodhi tree, is venerated by over 1.66 billion people round the world, primarily Hindu (1.161 billion) or Buddhist (0.506 billion) adherents. The leaves of the sacred fig have a distinctive heart shape with a pointed drip tip. Images of the tree appear as early as 2700 BC on pottery found at the Mundigak site near Kandahar, in Afghanistan. The species is native to the Indian subcontinent and Indochina.

Hindus circumambulate sacred fig trees as an act of worship called *pradakshina*. For Buddhists, Gautama Buddha attained enlightenment while meditating underneath a sacred fig, sometime around 500 BC. This original tree is now dead, but cuttings and seeds from its descendants have been spread globally in tropical and subtropical climes, its dispersal stimulated by its religious significance. The species' biology and broad tolerance of different conditions mark it as a possible invasive species.

↓ Painted goblet with a sacred fig leaf motif (period IV, c. 2700 BC), from Mundigak, an archeological site in Afghanistan, now located at the *Musée Guimet*, Paris.

↓ The Bharat Ratna medal in the form of a sacred fig leaf is the highest civilian award of India in recognition of exceptional merit in any field of human endeavor.

→ This sacred fig tree (*Ficus religiosa*) overwhelms the eastern gopura (ancient temple gateway) beside a tower located in the ruins of Ta Som, a late 12th-century Buddhist temple in Cambodia.

HEARTWOOD, SAPWOOD, AND THE BRITISH LONGBOW

Woodworkers often quiz one another on the best wood for a given purpose. What is the best wood for axe handles? Hickory (*Carya* species). For oars? White ash (*Fraxinus americana*). For the blades of old-fashioned ceiling fans? Cucumber magnolia (*Magnolia acuminata*).

THE STORY OF ÖTZI

The accumulation and deployment of knowledge of the best wood for which purpose has a deep history. How deep? In 1991, Ötzi, a natural mummy, was discovered high in the Ötztal Alps on the Austrian–Italian border. Born around 3275 BC, his body was found where he was murdered in around 3230 BC. Ötzi and his possessions offer a unique documentation of life in the European Chalcolithic (Copper) Age. Among other objects was an unfinished longbow (6 ft/1.82 m long) made of common yew (*Taxus baccata*). Ötzi's longbow was the same length and species of wood as the effective English longbow used in

---

THE LONGBOW

The nearly white sapwood of common yew can withstand tension; its golden, orange-brown heartwood withstands and releases from compression. A longbow's wood in cross section was shaped like the letter D and had the stretchy sapwood on the flat outside of the bow and the compressible, springy heartwood on the curved inside. The result is an alignment of wood properties that nowadays is created from composites of different woods held together by modern glues. Ötzi's longbow "kit" represents a technologically sophisticated use of wood from the end of the Stone Age using deep experience to select wood with unique properties, instead of glue technology.

---

the Hundred Years' War between England and France (1337–1453), some 4.5 millennia later. An even older longbow with a calibrated radiocarbon date between 4040 and 3640 BC is on display in The National Museum of Scotland.

COMMON YEW: RELIGION AND WAR

Likely from pre-Christian practice, common yews are traditionally planted in churchyards throughout Britain and Ireland, perhaps because of the symbolic implications of their dark, evergreen foliage. Robert the Bruce's forces used longbows made from the sacred yews at the Ardchattan Priory, in Argyll, in the Scots' victorious battle at Bannockburn against the English in 1314. The hail of arrows, numbered in the hundreds of thousands, shot from English longbows at the Battle of Agincourt (1415), along with the French heavy calvary's difficulties with the muddy terrain, gave Henry V a celebrated victory. The Hundred Years' War between England and France greatly depleted the supply of longbow yew across Europe and the churchyard yews may have represented a potential but small reserve supply. The replacement of longbows by firearms during the reign of Elizabeth 1 would have greatly diminished this demand.

THE AMERICAN REVOLUTION

Academic Thomas Esper reports that Benjamin Franklin argued for the use of longbows on the eve of the American Revolution in 1776 on the grounds that longbows were more accurate than muskets, could be shot four times in the interval of a musket's firing and reloading, any arrow wound is disabling, and bows and arrows can be resupplied more easily than muskets. Esper opined that central to the demise of the longbow and its replacement by muskets was the eventual abandonment of the organized training and lifelong practice required of the longbowman.

↓ In North America, the Osage orange (*Maclura pomifera*) has a superior wood for making bows. Early French settlers called the tree *bois d'arc* (bow-wood) as a result. The Osage Native Americans traded the wood over hundreds of miles and the bows were used effectively in warfare by the Comanches.

THE KULA RING

The Kula Ring is an ancient ceremonial exchange system involving thousands of individuals from 18 island communities in the Massim archipelago, which includes the Trobriand Islands and an associated large ring of other islands, in the Solomon Sea.

KULA JEWELRY EXCHANGE

Participants travel hundreds of miles of open ocean aboard outrigger canoes to exchange jewelry, which is a unique and unifying aspect of their culture. The broad rules of this trade are:

- Red shell-disc necklaces are traded to the north (circling the Kula Ring of islands in a clockwise direction).
- White shell armbands are traded to the south (counterclockwise.)
- If an opening gift is an armband, then the closing gift must be a necklace, and vice versa.

The exchange of Kula ritual valuables may be accompanied by trade through barter of commodities and utilitarian items.

OUTRIGGER CANOES

Outrigger canoes are logistically necessary for Kula Ring trading. Their construction requires a deep knowledge of wood, wood properties, and the best sources of the timbers needed. F. H. Damon, Professor of Anthropology at the University of Virginia, made extensive observations of the construction of these craft, noting: "Every boat, therefore, represents a division of labour expressed between building and carving, sailing, and horticultural expertise." Different parts, including the *Pandanus* leaf-fiber sails, the masts, and the outrigger float, and its connections to the rest of the canoe, are sourced from particular trees, often from specific locations.

BUILDING AND SAILING THE CANOES

The straight-grained masts used for the canoes come from a specific region on Muyua (Woodlark Island), where the soils and mid-length, slash-and-burn farming cycles produce superior masts. Land-use practices are reflected in the quality of the masts. Made from heartwood only, the keel is selected from tree species with strong, cross-grained wood at forest edges. Keel-wood is selected according to the natural arc of the tree's trunk, with the keel's size profoundly influencing the craft's other dimensions.

When tacking, the sail is moved toward the back of the outrigger—the boat's front then becomes the back—and the keel must maintain its integrity under the stress of this maneuver. A boat has two identical prow-boards, front and back. As described by anthropologist Bronisław Malinowski in 1922, prow-board installation involves the mwasila ceremony to rectify spirit, personification, and/or gender differences among the boat's components.

↑ A *Pandanus* sail being set as a Kula canoe is launched, at Kitava, in the Trobriand Islands, Papua New Guinea.

INDIGENOUS KNOWLEDGE AND BIOLOGICAL HERITAGE

The understanding of ecology, wood technology, sailing, and so on required for the Kula Ring ceremony (see page 134) reveals indigenous knowledge. The wisdom needed to construct sail-propelled canoes with outriggers and use them to navigate the ocean by reading the stars and waves is truly impressive. So is the courage needed to apply this knowledge to sailing the Pacific Ocean. Even if one could acquire such knowledge and skill, the biological heritage—the specific trees used for vessel construction—are unique to their island source.

THE WORLD OF "BIO-PIRACY"

The British explorer Sir Henry Alexander Wickham (1846–1928) transported 70,000 *Hevea brasiliensis* seeds from Santarém, in Pará province, Brazil, to London's Kew Royal Botanical Gardens in 1876. His intention was to break the Brazilian monopoly on rubber by using the seeds to establish rubber plantations in Southeast Asia. The seeds were propagated at Kew and plants were shipped to Sri Lanka, Malaysia, Indonesia, and other tropical destinations.

In Asia, *Hevea brasiliensis* can be grown in plantations. In its native Brazil, endemic diseases (notably South American leaf blight caused by the fungus *Microcyclus ulei*, along with other diseases and insects)

THE RUBBER TREE

The rubber tree (*Hevea brasiliensis*) reveals its biological heritage in its scientific name *brasiliensis* (meaning "of or from Brazil") and is the primary source of natural rubber. In 2022, over 90 percent of the world's supply of natural rubber came from Asia and less than 2 percent from Brazil. Given that the tree once only grew in the Amazon rainforest, its performance as an economically valuable plant has clearly expanded well beyond its range.

QUININE

Quinine comes from the bark of *Cinchona officinalis* from the tropical Andean forests. The fever-reducing properties of cinchona bark were known in South American cultures prior to European contact, and it long served as a malaria cure. Currently, knowledge of traditional medicines continues to be transferred to modern medicine. A knowledge of the properties of relatively unknown plants is valuable, but the charge for acquiring this knowledge is not always levied.

prevent successful plantations. Instead, New World rubber production involves rubber tappers locating widely dispersed *Hevea* trees in the rainforest and scoring the bark to collect latex for rubber. It is a labor-intensive process. Asian rubber plantations are much more productive per unit area or unit of labor than the efforts of solitary rubber tappers. This difference broke Brazil's monopoly on global production and led to economic collapse. Knighted in 1920 for his "services in connection with the rubber plantation industry in the Far East," Sir Wickham is considered a "bio-pirate" in Brazil.

Bio-piracy did not originate with Wickham, however. In 552, silkworm moth (*Bombyx mori*) eggs and their food, white mulberry (*Morus alba*), were smuggled from China to Constantinople (now Istanbul) in a move supported by Emperor Justinian, allowing for silk production in Byzantium. Domesticated plants grown by indigenous peoples became a significant product of the European "Voyages of Discovery," as is still the case today. The introductions of maize, potatoes, tomatoes, sunflower, avocado, several beans, cassava, chili peppers, quinoa, and so on were all the result of plants being domesticated by peoples of the "New World."

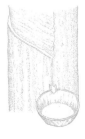

↓ Tapping a wild rubber tree in the Amazon. The bark is scored to promote the flow of latex-rich sap, which is collected in cups for about six hours.

THE TALKING TREE

Commonly called the Indaba tree, the bushveld cherry or the jacket plum, *Pappea capensis* is a small tree with a tasty, juicy fruit. *Indaba* is a term from the Zulu and Xhosa languages for a meeting to discuss important tribal business. King Lobengula of the Ndebele tribe (ca. 1836–1894) installed Indaba trees in his provincial capitals, now in Bulawayo Province, Zimbabwe. Many of these remain. In one case, after an Indaba tree died—becoming the victim of a strangler fig—the surviving fig tree grew and is now referred to as an Indaba tree. In general, the Indaba is a tree with a history as a site for important discussions to take place. Zimbabwean academic and author Dr. Yvonne Maphosa points out that in a traditional sense the cool shade of the tree might serve to cool the heat of arguments.

↓ King Lobengula of the Ndebele planted *Pappea capensis* trees to create meeting places to discuss community concerns in all his capitals. The trees and the meetings became known as *Indaba*, a Zulu word that has become part of South African English.

→ Like the related Chinese lychee (*Litchi chinensis*) tree, the fruit of the Indaba tree is sweet, tasty, and juicy. It has traditionally been considered a medicinal plant for infections. Recent studies from the University of KwaZulu-Natal have isolated and identified antimicrobial, antioxidant, and cytotoxic compounds in the plant's leaves, which supports the traditional use of the plant.

THE DIDGERIDOO

T he digeridoo is a wind instrument typically made from a hollow tree or, in some cases, a hollow tree limb. It can also be made of bamboo or PVC pipe. It is played by vibrating the lips, which are pressed against the smaller end of the tubular instrument. It does not have a mouthpiece; instead, a rim of bee's wax at the playing end of the instrument promotes a seal. The instrument is cylindrical or occasionally conical. Typically, didgeridoos are around 4 ft (1.2 m) long, but the instruments can measure 3–10 ft (1–3 m) in length. The longer the instrument, the lower its pitch or key.

MAKING A DIDGERIDOO

A traditional didgeridoo uses the Australian landscape and its processes when sourcing the material needed to make the instrument. One important stage is identifying areas of high termite activity, as this hollows out the tree and makes it suitable for crafting a didgeridoo. When termites gain entry to *Eucalyptus* trees (one of the types of trees used to make the instrument), the trees' outer sapwood chemically repels the termite. The termites only feed on the heartwood. Knocking or thumping the hollow trees gives an indication of their hollowness, resonance, and the potential tone of the resultant instrument. When a suitable tree is identified, it is harvested, debarked, cleaned out, and cut to length.

PLAYING A DIDGERIDOO

A technique called circular breathing carries the didgeridoo's drone continuously for as long as 40 to 50 minutes. Circular breathing involves breathing in through the nose while simultaneously using the muscles of the cheeks to compress the cheeks and release air stored in the mouth. Breathing out then refills the air stored in the mouth. Using circular breathing, one can constantly expel the air from the mouth while filling and emptying the lungs.

DIDGERIDOO ORIGINS

In the Kakadu National Park region of Northern Australia, the didgeridoo seems to have been in use for less than a thousand years based on archeological dating of rock art paintings. The name didgeridoo does not seem to have an Aboriginal Australian origin and instead appears to be derived from the instrument's onomatopoeic sound. It would seem to be a relatively recent innovation in the continuing context of a very ancient human culture.

↓ The didgeridoo is a wind instrument played by vibrating one's lips while using a special breathing technique in order to produce the familiar continuous droning sound.

LARGEST AND OLDEST LIVING BEINGS

Aside from clonal organisms, including clonal trees, trees generally hold all the records in terms of achieving great heights and weights, as well as remarkable ages.

THE WORLD'S LARGEST TREES

The world's largest tree, a giant sequoia called General Sherman (*Sequoiadendron giganteum*), weighs 1,235 US tons (1.12 million kg) and is sixteen times larger than the blue whale. The tallest tree is a coast redwood (*Sequoia sempervirens*), called Hyperion, at 380.3 ft (115.92 m), taller than the Statue of Liberty (305 ft/93 m). The record diameter belongs to a Montezuma cypress (*Taxodium mucronatum*) in Mexico, which has a diameter of 37.73 ft (11.5 m) above the root

AGE AND HEIGHT

The largest trees are not always the oldest. As Liu and colleagues reported, the largest trees grow in a range of environments, but the oldest are mostly found on steep, rocky, cold, arid, or less fertile sites, especially in the western parts of China and North America. These conditions lead to slow growth rates, a trait often correlated with longevity (see page 104). They also tend to be places that have escaped direct human impacts.

swell. The tree with the largest crown spread is a multi-trunk banyan (590.55 ft/180 m across). Banyans are unusual in producing new trunks from prop roots and thereby spreading widely (see page 56). The highest ranked single-stemmed tree is a eucalyptus with a crown measuring 239.5 ft (73 m) across.

Clonal "trees" attain even larger sizes and ages than single-stemmed trees. For example, the Pando aspen clone (see page 57) covers an area 36 times larger than the banyan record holder, is 6 times heavier than the giant sequoia, and, at 14,000 years old, is 3 times older than the oldest tree. The individual aspen stems reach only 130 years.

THE WORLD'S OLDEST TREES

Aside from clonal organisms and those that have demonstrated exceptional dormancy—some even reviving after burial in glacial ice—trees are the planet's oldest living beings. While the largest trees are an even mix of gymnosperms and angiosperms, Dr. Jiajia Liu, of Fudan University in China, and colleagues, found that 28 of the 29 trees over 2,000 years old are gymnosperms. The oldest single-stemmed tree is a bristlecone pine (4,854 years); a giant sequoia is second (3,266 years). The oldest angiosperm is a baobab in Zimbabwe (2,419 years).

← Mexico's national tree, the Montezuma cypress, is found in moist to seasonally flooded soils. The largest grows on church grounds in Santa María del Thule.

SEEDS IN SEARCH OF SAFE SITES

Henry David Thoreau wrote: "convince me that you have a seed there, and I am prepared to expect wonders." The giant sequoia tree is just such a wonder. The tiny seeds, each weighing less than one-tenth of a gram, carry the DNA blueprint that produces the Earth's largest trees, those that will grow to be over 50 billion times heavier than the seeds from which they came.

SURVIVING THE ODDS

The establishment of new seedlings is a critical life stage for all trees, yet the odds are steep. Thousands or even millions of seeds are produced for each seed that eventually becomes a mature tree. How do tree species deal with those difficult odds? Two British ecologists, John Harper and Peter Grubb, developed a conceptual approach to answer that question. Harper defined "safe sites" as the conditions, at the scale of individual seeds, that are required for successful germination and establishment. Grubb defined the "regeneration niche" as the requirements for successful reproduction, dispersal, germination, and seedling establishment—in other words, the ways that different species reach safe sites. Different species of trees have different kinds of safe sites, and they reach those safe sites in different ways.

PIN CHERRY

The pin cherry (*Prunus pensylvanica*) is a small, short-lived tree with seeds that can remain dormant in the soil seed bank for decades to centuries until they are triggered to germinate by increased light, warmth, or soil nitrate levels following disturbance to the forest canopy. If cottonwood seeds are in "search" of the right patches for establishment, the pin cherry is in search of the right time to germinate.

The simplest case occurs when seedlings become established below mature trees following annual seed crops. Species that germinate and persist in dense forests are called "shade tolerant." However, the individuals that establish in this way usually require future canopy disturbances in order to enter the canopy themselves. In many other cases, seedlings are absent below canopy trees. Instead, safe sites for these species are separated from the seed-bearing trees in space or in time or both.

~ Seeds seeking the right place ~

Meandering rivers continually erode some banks and deposit alluvial soil on others. Cottonwoods (*Populus* species) have small, wind-dispersed seeds that colonize the fresh alluvial soils (see Chapter 9, page 106) at some distance from the seed-bearing trees. Species with this dynamic are called "fugitive" species, since they are always fleeing to new sites and disappearing from old ones.

~ Seeds seeking the right time ~

A Costa Rican species of *Tachigali*, called the "suicide tree," creates its own safe sites. *Tachigali versicolor* reaches a large size, flowers once (a trait called "monocarpy"), and then dies, creating a sunlit gap just when its own seeds are in large supply.

~ Seeds seeking the right place and time ~

Giant sequoia crowns hold around 30 years of unopened cones, each with hundreds of seeds. The heat of a fire is required to melt the resins that seal these cones shut. The cones typically open slowly enough after fire that hundreds of thousands of seeds are released after the passage of the flames (see page 146). The safe sites for these seeds are patches of open mineral soil created by fires that consume the surface leaf litter.

→ The cone of the giant sequoia is about 2–3 in (5–7.5 cm) long. The tiny seeds are released when the cones open, as shown here.

LIVING WITH FIRE

Long evolutionary exposure to fires, many years before humans began using fire, has resulted in adaptations that allow trees to tolerate, exploit, and even depend upon fire (see Chapters 6 and 8).

One of the most remarkable fire adaptations is called "serotiny," the holding on to seeds until fire triggers their release. For instance, serotinous cones in gymnosperms (such as the giant sequoia and several pine species) and serotinous fruits in angiosperms (such as proteas and eucalyptus) are sealed shut by resins. In the intervals between fires, these reproductive structures accumulate on tree branches, allowing the simultaneous release of many years of seed production after fire. The seeds stored in the canopy are the "canopy seed bank." The fires that result in cone opening may also consume the thick leaf litter on soil surfaces, so the released seeds find ideal conditions for germination. The seeds of some species are also stimulated to germinate by chemicals contained in smoke.

↓ Table mountain pine (*Pinus pungens*) from the eastern USA is found on fire-prone mountain ridges with dry soils due to rapid drainage.

↓ First-year seedlings of table mountain pine establishing in a recent fire scar in the spring following a fire.

→ Most table mountain pine cones open after exposure to fire, though this species, like several others, "hedges its bets" with a few cones opening after exposure to warm, dry conditions. Dr. Ed Johnson and colleagues from the University of Calgary investigated serotiny in another pine, *Pinus contorta*, documenting the ideal heat exposure required to open the cones.

DROUGHT AND THE BAOBAB

Trees are limited at high latitudes and elevations by short growing seasons and cold temperatures. In warmer climates, trees are limited in areas of low precipitation, on the one hand, or constant inundation on the other. Chemistry can be a factor, too, including salty coastlines (see page 150) and barrens caused by nutrient deficiencies.

Baobabs (*Adansonia*) occur on sites that are at the very limit of moisture availability for trees. The genus consists of eight species, six endemic to Madagascar and one each in Africa and Australia. One of the oddest and most distinctive of all trees, baobabs are pachycauls (trees with unusually stout trunks for their height) and have few branches. According to one legend, the gods felt the tree was too proud, perhaps because of its grand stature or the many ways people used it, so they plucked it from the ground and planted it with its roots in the air, hence the common name, "upside-down tree."

BAOBAB ADAPTATIONS

Baobabs have many adaptations for drought. The leaves are deciduous in response to eight- to nine-month dry seasons. Rainfall is rarely high enough to percolate into the soil, but baobab bark is spongy and absorbs water directly. Reportedly, the trunk can store over 26,400 US gallons (100,000 liters) of water, which increases the diameter by 2.5 percent in the wet season. During exceptional drought, elephants tear apart the trunks to reach this water supply. The tree shows fire adaptations, too. After fire, baobabs can sprout at the base and often form rings of clustered trunks. The seeds, dispersed by elephants and other large animals, have a long dormancy, germinating after fire or following passage through animal digestive systems. The flowers are adapted to bat pollination, opening in late afternoon and into the evening.

↑ The baobob tree, here in Madagascar, is one of the Earth's most distinctive tree species. It is found in very dry habitats that are mostly dominated by shrubs and non-woody species, and has many adaptations to drought.

BAOBAB KEY FACTS

Despite growing in extremely dry environments, baobabs are among the largest angiosperm trees, reaching diameters of over 36 ft (11 m) and with crown spreads over 164 ft (50 m). To provide fire protection, the bark is thick (greater than 8 in/20 cm) and fire-resistant. The trees lack easily distinguished growth rings, but carbon dating has shown that they reach ages in excess of 2,000 years.

LIVING WITH SALT AND STORMS

Tropical coastlines, with their toxic levels of salt and the constant movement of the tides, may seem an unlikely environment for trees and yet this is where we find a group of some 50 tree species, collectively called "mangroves," which have a series of remarkable adaptations for this extreme environment. These species come from 16 families, suggesting that adaptations have occurred independently.

To deal with salinity, mangroves have thick cuticles on surfaces to exclude salt. They also shed older leaves and use metabolic pumps to remove salt from living tissues. To deal with low oxygen in the root environment, adaptations include pneumatophores (aerial roots adapted for gas exchange), aerenchyma cells in stems that promote diffusion of oxygen, and prop roots that increase surface area for oxygen absorption. To establish seedlings quickly in submerged areas, some mangroves even have seeds with "live birth"; that is, seeds that germinate while still attached to the parent tree.

↓ Mangroves have several strategies for salt exposure, including a thick, waxy covering for the leaves and metabolic pumps that excrete salt.

↓ The red mangrove (*Rhizophora mangle*), which grows throughout the tropics, has stilt roots that become exposed at low tide.

→ The red mangrove has seeds that germinate while still attached to the parent tree. In this species, the developing seedling has a long, dagger-like axis, allowing it to quickly reach the underwater sediments for rooting, while simultaneously holding its developing stem within easy reach of the water surface and atmosphere above.

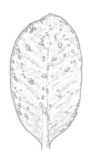

PLAYING DEFENSE, TALKING TREES, AND MARAUDING ANTS

Trees have nutrient-rich tissues and are the basis of the energy flow in forested ecosystems. But they are also rooted in place for decades to millennia. To protect themselves from enemies, they have evolved an astounding array of defensive strategies.

DEFENSE STRATEGIES

It may be counterintuitive, but defense is most cost-effective when resource levels are low. Where resources and growth rates are high, a tree can replace any losses to herbivores quickly; this is not the case where resources are low. Indeed, Dr. Edwin Pos and colleagues, of Utrecht University, found that "strategies to deter herbivores (e.g. latex, resin) become abundant across the large expanses of Amazonia where soils are nutrient poor." The latex produced by one of those trees, *Hevea brasiliensis*, is the source of natural rubber (see page 136).

As lepidopterist Dave Wagner remarked, you can thank tropical caterpillars for your next car or bus trip (because rubber trees produce latex as a defense against them). You should also thank plant defense for your next caffeinated beverage. The natural world is a library of repellant and toxic chemicals that have evolved over millions of years to interact with animal physiological processes. Some are poisonous to us, but others are used as spices, flavorings, and medicines.

INDUCIBLE DEFENSES

Inducible defenses show trees can distinguish enemy attack from mechanical damage, a trait that might be called a sense of "smell," and that they can communicate. Trees under attack can produce chemical signals that warn other trees nearby to deploy their defenses. Such signals are sent through the air as volatile compounds or via root systems and the mycorrhizal fungi that connect trees.

Some defensive strategies are called constitutive because they are permanent and hard-wired into each tree. Others are inducible and are deployed when the tree is under attack. If enemies are always present, it may be more cost-effective to have constitutive defenses, but where they are variable, inducible might be better.

~ Constitutive defense ~

This type of defense begins with hard surfaces and coverings. Leaves and young stems are covered with thick, waxy cuticles or dense, stiff hairs. Some species incorporate abrasive silicon particles or calcium oxalate crystals in their leaves. Bark is fibrous and rich in hard-to-digest lignin. Stems can be covered with thorns or prickles—for some species, injury results in sprouts that are more densely covered with these than the pre-injury stems.

DEFENSIVE MUTUALISMS

Coevolution has produced some amazing defensive mutualisms. There are two types. Some trees produce volatile chemicals that attract the enemy of the enemy, such as when trees attract wasps that parasitize caterpillars. Some defensive mutualisms even involve attracting, housing, and feeding ants, which then act as the tree's sentinels and defenders. Insects can be nourished through extrafloral nectaries. Housing is provided as "domatia," such as the hollow thorns of acacias and the hollow stems of *Cecropia*.

↓ *Azteca* ants patrol a *Cecropia* stem to remove insect enemies. In return, the tree provides their nesting and feeding sites.

CHEMICAL TREE ATTACK

Trees may also produce chemicals, called allelochemicals, that sabotage competing trees. One theory is that such chemicals affect the mutualistic mycorrhizal fungi of the competitors. Temperate walnut trees are a prime example, and the understory of walnut stands are more open and less diverse in tree species than nearby forests that lack walnuts. The myriad defensive strategies of trees often involves communication between trees, a subject of great interest in ongoing research.

GLOSSARY

aerenchyma
A soft tissue in the leaves, stems, or roots of some plant species, which has abundant air spaces to facilitate the movement of gases, especially the oxygen essential to all living tissues.

allometry
Regular proportions in the sizes and shapes of tree tissues and structures (e.g., tree canopy width and trunk diameter). Also refers to comparable regularities in the geometry in forest stands.

angiosperm
Literally, covered seed. The angiosperms comprise the flowering plants, defined by the structure of the female reproductive organs in which the ovules (and later the developing seeds) are contained within ovaries or parts thereof (the carpels). Angiosperm trees are known collectively as hardwoods.

anoxic
The condition of being depleted of the oxygen on which living tissues depend (usually in reference to fresh or salt water).

cambium
A living zone of undifferentiated tissues that produces secondary growth in woody plant stems through the production of xylem and phloem, or in the case of the cork cambium, the bark. Also called the lateral meristem.

chloroplast
The organelles (small organs) that carry out photosynthesis in plant cells.

chronosequence
A collection of descriptions of developing forests arranged in a temporal sequence.

dicotyledon
Often abbreviated to dicot, one of two major divisions of flowering plants in which the embryo has two seed leaves, structures that often contain stored carbohydrates for germination and early development.

endosperm
The component of the seed that contains stored carbohydrates and other nutrients for seed germination and early development.

epicormic stems
New stems produced by growth from long dormant buds found in the trunk and branches, often as the result of injury, including exposure to fire.

gymnosperm
Literally, naked seed. Seed plants that lack true flowers, the seeds developing without the covering of an ovary. Many gymnosperms are familiar as the cone-bearing, needle-leaved pines, spruces, and firs.

heartwood
In most trees, the older xylem (wood) found in the tree's center is chemically transformed to form heartwood, which is often darker, heavier, more decay-resistant, and stronger than the younger sapwood surrounding it.

inflorescence
The distinctive arrangements of flowers on different plant species, including spikes, corymbs, racemes, and heads.

lenticel
Porous tissues on plant stems that allow the exchange of gases, permitting oxygen to reach the living inner tissues.

lignin
Complex organic chemicals produced in the cell walls of vascular plants, which bind cellulose fibers and provide stiffness.

meristems
Zones of undifferentiated tissues that produce either primary growth (the apical meristems) or secondary growth (the lateral meristem or cambium).

monocotyledon
Often abbreviated to monocot, one of two major divisions of flowering plants in which the embryo has a single seed leaf, a structure that often contains carbohydrates for seed germination and early development.

monopodial
Plants that grow upward with a single stem.

mycorrhiza (pl. mycorrhizae)
Symbiotic fungus that grows within the cells of plants.

phloem
Complex tissue in vascular plants with a primary function of transporting the sugars made by photosynthesis. Phloem tissues may also contain specialized cells that provide mechanical support, flexibility, storage, and protection from herbivores.

photosynthate
A substance produced by photosynthesis (e.g., sugar).

photosynthesis
The process by which plants use solar energy to convert water and carbon dioxide to sugars that are then used as the plant's energy source for growth and reproduction.

pit vessel
In angiosperms, pit vessels are specialized structures in the sapwood that allow for greater rates of water transport than found in gymnosperms.

pneumatophore
Aerial root adapted for gas transport, especially advantageous to plants growing in anoxic conditions.

sapwood
Wood surrounding the heartwood and involved in transporting water. Typically, lighter in color than heartwood.

stomate (pl. stomata)
Pore in plant leaves that serves as the entry point and exit point for water vapor lost through evapotranspiration. Guard cells control the opening and closing of stomates as a function of environmental conditions.

tracheid
Long, tubular, pitted cell in the xylem of trees.

transpiration
The loss of water from plant tissues through diffusion of water from the stomata.

xylem
Derived from the Greek word *xylon*, meaning "wood," xylem is vascular plant -tissue that conducts water, along with dissolved elements, upward from the roots and also provides mechanical support to the plant.

FURTHER READING

Readers may type reference information into Google Scholar (scholar.google.com) for more information and often a copy of the reference.

Chapter 1: Niklas, K. J. 2007. *Tree Physiology* 27, pp433–440.

Shugart, H. H., P. White, S. Saatchi, and J. Chave. 2022. *The World Atlas of Trees and Forests.* Princeton University Press. Princeton. New Jersey.

Stein, W. E., C. M. Berry, J. L. Morris, ..., C. H. Wellman, D. J. Beerling, and J. R. Leake. 2020. *Current Biology* 30: pp421–431.

Chapter 2: Gatti, R. C., et al. 2022. The number of tree species on earth. PNAS.

Hubbell, S. P. 2015. Estimating the global number of tropical tree species, and Fisher's paradox. PNAS.

Janzen, D. H. 1970. *American Naturalist* 104: pp501–528.

Ricklefs, R. E., and D. S. Schluter. 1994. *Species Diversity in Ecological Communities.* University of Chicago Press.

Slik, J. W. F., et al. 2015. An estimate of the number of tropical tree species. PNAS

Chapter 3: Baker-Brosh, K. F., and R. K. Peet. 1997. *Ecology* 78 (4): pp1250-1255.

Hershey, D. R. 1991. *The American Biology Teacher* 53: pp458–460.

Peppe, D. J., et al. 2011. *New Phytologist* 190: pp724–739.

Venturas, M. D., J. S. Sperry, and U. G. Hacke. 2017. Plant xylem hydraulics: what we understand, current research, and future challenges. *Journal of Integrative Plant Biology.*

Chapter 4: Carlton, W. R. 1939. *The New England Quarterly* 12: pp4–18.

Chave, J., D. Coomes, S. Jansen, S. L. Lewis, N. G. Swenson, and A. E. Zanne, 2009. *Ecology Letters* 12: pp351–366.

King, D. A. 1990. *American Naturalist.* 135: pp809–828.

Lefsky, M. A. 2010. *Geophysical Research Letters* 37, L15401.

Chapter 5: Chomicki, G., M. Coiro, and S. S. Renner. 2017. *Annals of Botany* 120: pp855–891.

Hara, T. 1984. *Annals of Botany* 53: pp181–188.

Horn, H. S. 1971. *The Adaptive Geometry of Trees.* Princeton Press. Princeton. New Jersey.

Roth-Nebelsick, A., T. Miranda, M. Ebner, et al. 2021. *Palaeobiodiversity and Palaeoenvironments* 101, pp267–284.

Chapter 6: Crisp, M. D., G. E. Burrows, L. G. Cook, A. H. Thornhill, and D. M. J. S. Bowman. 2010. Flammable biomes dominated by eucalypts originated at the Cretaceous-Palaeogene boundary. *Nature Communications.*

Wojtech, M. 2020. *Bark: A Field Guide to Trees of the Northeast.* Brandeis University Press.

Chapter 7: Eissenstat, D. M. and R. D. Yanai. 1997. *Advances in Ecological Research* 27: pp1–60.

Ma, Z., Guo, D., Xu, X. et al. 2018. *Nature* 555, pp94–97.

Remy, W., T. N. Taylor, H. Hass, and H. Kerp. 1994. *Proc Natl Acad Sci USA.* 1994 Dec 6; 91(25): 11841–3.

Smith, S. E., and F. A. Smith. 2011. *Annual Review of Plant Biology* 62: pp227–250.

Chapter 8: Grime, J. P. 2001. *Plant Strategies, Vegetation Processes, and Ecosystem Properties.* John Wiley and Sons. Hoboken. New Jersey.

Janzen, D., and P. S. Martin. 1982. *Science* 215: pp19–27.

Marks, P. 1974. *Ecological Monographs* 44: pp73–88.

Reich, P. B. 2014. Journal of Ecology 102: pp275–301.

Stettler, R. 2009. *Cottonwood and the River of Time.* Washington University Press.

Valenta, K., and O. Nevo. 2020. The dispersal syndrome hypothesis: how animals shaped fruit traits and how they did not. *Functional Ecology.*

Chapter 9: Reiners, W. A., and G. E. Lang. 1979. *Ecology.* 60: pp403–417.

Shugart, H. H. 1984. *A Theory of Forest Dynamics: The Ecological Implications of Forest Succession Models.* Springer-Verlag. New York.

Walker, L. R., J. Walker, and R. J. Hobbs. 2007. *Linking Restoration and Ecological Succession.* Springer-Verlag. New York.

Watt, A. S. 1947. *Journal of Ecology* 35: pp1–22.

Chapter 10: Botanic Gardens Conservation International & Fauna & Flora International. Securing a Future for the World's Threatened Trees—A Global Challenge (BGCI/FFI, 2021).

Botanic Gardens Conservation International. State of the World's Trees (BGCI, 2021).

FAO. 2022. In Brief to The State of the World's Forests 2022. Forest pathways for green recovery and building inclusive, resilient, and sustainable economies. Rome, FAO.

Terborgh, J., et al. 2001. *Science* 294: pp1923–1926.

Chapter 11: Anon.2023. "Traditional Pacific Island Crops" https://guides. library.manoa.hawaii. edu/paccrops

Damon, F. H. 2017. Trees, Knots, and Outriggers: *Environmental Knowledge of the Northeast Kula Ring.* Berghahn Books, New York and Oxford.

Esper, T. 1965. *Technology and Culture* 6: pp382–393.

Jackson, J. 2008. *The Thief at the End of the World: Rubber, Power, and the Seeds of Empire.* Viking. New York.

Chapter 12: Grubb, P. J. 1977. *Biological Reviews,* 52, pp107–145.

Harper, J. 1977. *Population Biology of Plants.* Academic Press. Cambridge. Massachusetts.

Johnson, E. A., and S. L. Gutsell. 1993. *Journal of Vegetation Science* 4: pp745–750.

Liu, J., S. Xia, D. Zeng, C. Liu, Y. Li, W. Yang, B. Yang, J. Zhang, F. Slik, and D. B. Lindenmayer. 2021. Age and spatial distribution of the world's oldest trees. *Conservation Biology.*

Pos, E., et al. 2023. Unraveling Amazon tree community assembly using maximum information entropy: a quantitative analysis of tropical forest ecology. *Scientific Reports.*

INDEX

ACKNOWLEDGMENTS

We appreciate the help of Ruth Patrick and the rest of the excellent team at UniPress: Lindsey Johns, Tugce Okay, Ian Durneen, Caroline West, and Robin Pridy. We thank Sassan Saatchi and Jérôme Chave, our co-authors on an earlier book, for continuing discussions on trees and forest ecosystems. Herman Shugart is eternally grateful for a lifetime of collaboration with his graduate students—you have taught me so much. Peter White thanks students and colleagues for their enthusiasm and tireless efforts to refine our understanding of the function and importance of trees and to surprise us with new findings. Last, and certainly not least, we thank Ramona Shugart and Carolyn White, for essential support as we have pursued an understanding of our forested world.

ABOUT THE AUTHORS

Herman Shugart holds the W. W. Corcoran Chair in Environmental Sciences (Emeritus) at the University of Virginia and has produced more than 490 scientific publications that largely involve systems ecology and ecosystems modeling, strongly focused on landscape and regional and global change. His book *How the Earthquake Bird Got Its Name and Other Tales of Unbalanced Nature* (Yale University Press) is considered a classic in modern ecology.

Peter White is a plant ecologist whose research interests have focused on the ecology of disturbances ecology, patterns of biological diversity, and conservation. For 28 years, he directed the North Carolina Botanical Garden (University of North Carolina at Chapel Hill). The Garden's conservation programs received a Program Excellence Award from the American Public Garden Association. His book, *Wildflowers of the Smokies,* won First Prize for natural history books about national parks. Peter was named one of the 100 most influential people in the history of Great Smoky Mountains National Park.

In 2022, Shugart and White, along with Sassan Saatchi and Jerome Chave, published *The World Atlas of Trees and Forests* (Princeton University Press), winner of the Dartmouth Medal (American Library Association) as the outstanding reference book of that year.